Why Do I Have to Know Mathematics?

You Don't

But That's the Problem

By

Dr. Rodney E. McNair

ISBN: 1456588087
ISBN-13: 9781456588083
LCCN: 2011901676

Contents

Chapter One

The Big Question

Teacher: Why do you think you have to learn mathematics?
Student: I know that I have to be able to count and stuff, so I can count my money and not get cheated, but I don't know how I would ever use algebra.

It seems that at some point during our formal educations, we have all asked the same question. "Why do I have to learn mathematics?" The standard answers frequently given by teachers include things like, "mathematics is all around us," "you need mathematics to get a good job," or "you need this course so that you can take the next course." These answers may provide motivation for some students, but they do not provide much insight into the real importance of mathematics. As we will see later, although these and other common answers to the question are based on fact, they do not come close to explaining the real reasons why mathematics is so important and why you have to learn mathematics.

The question becomes even more salient to students who may see that a number of people who do not have significant mathematics skills beyond computation of basic fractions, interest, and percents live average and even exceptional lives. Often one need look no further than the students' parents to find an example of a successful individual who uses little mathematics. For all of the emphasis we place on mathematics in schools and in the media, we can get through an entire lifetime without factoring a polynomial, determining the cosine of a thirty-degree angle, or completing the square to find the equation of a circle.

Mathematics has always played a major role in scientific discovery. While this role is not often clear to junior high school and high school students, most students can accept the argument that careers in these fields require the advanced mathematical knowledge provided by formal high school and college classroom instruction. These arguments may motivate students who want to go to college and major in science. However, the larger number of students who are not interested in science careers are left to ponder the question: why do we have to learn this stuff? Those students who are not intending to go to college are even less motivated by the usefulness of mathematics in science and engineering.

College entrance requirements and exams stress high mathematics achievement. Colleges require students to study four years of mathematics in high school that usually includes some combination of algebra, geometry, calculus, and statistics, and many colleges encourage students to take advanced placement calculus courses in high school. Government and industry leaders warn us that the United States is lagging behind the rest of the world in mathematics achievement. The number of jobs that require mathematics to support higher levels of decision-making skills is increasing. The projected result is that jobs will be lost to foreign competition. Thus the cry goes out to increase mathematics achievement levels.

All of the efforts to raise the alarm may have an unintended backlash. The insistence that advanced mathematics skills are necessary for success together with the evidence that few people use mathematics (algebra, calculus, etc.) on a regular basis supports the characterizations of mathematics as a right of passage, mathematics as a sieve, or of mathematics as a gatekeeper. Unfortunately (or fortunately depending on your perspective), once you have passed through the gate, you can forget the gatekeeper.

Consistent with the mathematics as a gatekeeper to success viewpoint, many students treat mathematics as no more than an obstacle to overcome. They look forward to the day when they will never have to study or do mathematics again. Many high school and college graduates say that mathematics (beyond the basics) was the most useless thing they had to study in school. This suggests that many students complete their formal education without ever really finding out why mathematics was a part of their curriculum. They never find an answer to the question "Why do I have to study mathematics?"

Students start to study algebra at a point in their lives when their interests are expanding and they are becoming more aware of the world around them. This is a time when they should be open to lots of new ideas. So why not algebra? When these young people ask why they have to learn algebra, they really want to know how algebra (and other more advanced mathematics content) is relevant to their lives. Surely this is a simple question for any mathematics teacher, policymaker, or textbook author to answer, yet the question remains.

No one ever asks why they have to learn to read or write. Even students who do not like chemistry or physics understand the importance of science. So why do so many people question the need to study mathematics? The fact that so many have asked the question suggests that it is a legitimate question and it deserves a proper answer. Also, as we will see, there truly are circumstances and conditions in world today that make it critically important that everyone know and understand why mathematics is so important. My goal in this book is to provide you with a clear unambiguous answer to the question "Why do I have to learn mathematics?"

The Context

Before we get to the answer to the burning question, we need to place the question in its proper context. Jobs and national dominance in science may be important, but these are not the reasons that every child in school needs to learn mathematics. It is not too strong a statement to say that our children's and our nation's futures depend on raising mathematics achievement levels. But it is not jobs and dominance in science that are at stake. Why we have to know mathematics is the question that is at the heart of a national and international effort to raise mathematics achievement levels. This heightens the importance of our question in terms of understanding the national importance of mathematics achievement, in terms of understanding why current achievement levels are so low, and in terms of understanding how to improve performance.

Efficiency versus Effectiveness

First, we must understand that low achievement in mathematics is not the result of a broken school system. Our school systems have always produced a number of talented students who possessed sufficient mathematics skills to go on to become scientists and engineers, but our school systems have also always produced a larger number of students who do not have sufficient mathematics knowledge to pursue careers in science and engineering. It may even be reasonable to not expect that every child will learn high levels of mathematics. But how many will and what is the fate of those who do not?

Low achievement in mathematics is a natural consequence of a school system that was designed to be efficient at the cost of being effective. The efficiency of our school systems is based on getting the most students through the system for the least amount of money. Effectiveness of school systems should be measured by the academic achievements of all of the students in the system; however, in reality effectiveness is more a matter of whether or not the system provides a large enough pool of talented graduates to sustain our world dominance in business, science, and technology. This is based on a limited view that mathematics is for science and or to get a job.

The conflict between efficiency and effectiveness arises when higher achievement standards are set. To make the system more effective, we set higher achievement levels, but to reach the higher achievement levels, students need greater supports. Providing these supports cost more money. Taxpayers and elected officials are not willing to pay the extra money, perhaps because they do not truly understand what is at stake. As a result, the higher standards are not reached, and effectiveness loses out to cost efficiency.

The problem is not the teachers, the students, or the parents. All of these players have been caught up in the battle waged between efficiency and effectiveness. The results are finger pointing, name calling, and accusation slinging. The problem is not that students cannot do mathematics.

The problem is not that teachers don't teach mathematics or that parents don't care about their children's education. The problem is that our education system is not designed and/or funded at the level appropriate to teach mathematics to all students. Thus, much of the anxiety and stress that people feel over mathematics is not due to their own shortcomings, but are instead the manifestation of the war between efficiency and effectiveness.

Understanding why mathematics is so important is central to efforts to increase funding, improve efficiency, and increase effectiveness of schools. It is not just that students need to know why mathematics is important. Policymakers, curriculum developers, and teachers also need a clearer understand of why mathematics is important. Far too often, we hear vague arguments and statements about the role that mathematics plays in getting better jobs, brighter futures, and responsible citizenship. It is time to move beyond motivational platitudes and provide an answer to the question. Once the answer is provided we can help students to develop personal motivation for the study of mathematics.

I'm Not Very Good at Mathematics

In our society, it is acceptable to state that we are not good at mathematics. When we utter comments like "I am not very good at mathematics" or "Mathematics was never my best subject," what weakness are we confessing? Why is there no shame in such statements? It could be that because the reason for studying mathematics is not clearly known, no value is placed on being able to do mathematics or not being able to do mathematics. Since those who make such statements are not in the minority, they may feel no social stigma. In the past (and still in some cultures), knowledge of mathematics was equated with one's ability to think. If this were the case in our society, fewer people would be ready to admit a weakness in mathematics. As we will see, while knowledge of mathematics may be separated from the general ability to think, a lack of mathematics ability may limit what it is that you are able to think about.

Mathematics is not a way to think. We think analogically, deductively, and inductively, and mathematics may make use of all of these forms of thought. Mathematics is used to create abstract contexts for thinking. Doing mathematics supports reasoning by abstracting situations into manageable entities that can be manipulated by a well-developed set of procedures and laws. The need and choice to perform these abstractions is at the heart of why we need to know mathematics. We will see later why so few people choose to make and use these abstractions.

The disconnection between mathematics and our everyday lives at the heart of the question, "Why do I have to learn mathematics?" allows us to admit weakness in mathematics without shame. Since mathematics is only relevant to jobs and issues that we have not chosen to pursue, then we suffer no more embarrassment in admitting a weakness in mathematics than we do by admitting that we cannot bench press 300 lbs. From a health and conditioning perspective, we may not need to be able to bench press 300 pounds, but we should all probably be able to lift more weight than we currently can. Also, from a sociocultural perspective, we may not need a college degree in mathematics, but we should all be able to do more mathematics than we currently can.

The disconnection between mathematics and our routine lives is caused by and a contributor to the natural gap between the amount of mathematics that scientists and mathematicians know and the amount that the average person needs to know. There have always been a relatively small number of scientists and mathematicians who have worked on and solved problems and issues confronting society and have worked to uncover the mysteries of our planet and universe. This subgroup of intellectuals has seemingly always existed and benefited the larger society. They invented or discovered things like fire, the wheel, the printing press, light bulbs, telephones, penicillin, electricity, and computers.

In 1492 while many common folk may have believed that the world was flat, this small group of intellectual leaders had known that the world was round since the days of Eratosthenes around 250 B.C. In the past, it was sufficient for the common citizen to confine his or her attention to the flat regions in which they farmed, shopped, and lived. Hunting and gathering did not require advanced knowledge of mathematics, and neither did planting and farming. And with few exceptions, most jobs in the twentieth century did not require advanced knowledge of mathematics. Most people spent their time dealing with the day-to-day demands of family and work. With the pressing questions of personal and family well-being to be concerned with, people can and need to rely on the subgroup of intellectuals to deal with more complex issues. This adds to and widens the gap between mathematics and society. But because the subgroup has relied more and more on the use of mathematics to advance our science and society, mathematics has become more and more relevant to our personal and family well-being.

Over time the small intellectual groups have developed a great number of technologies to "simplify" and "improve" the lives of the masses. However, a result of these advancements is that the world is a much more complex and complicated place. Social, financial, and political systems have become much more complex. We face new environmental and health issues and can no longer rely on the small group of intellectuals to solve problems for us. We must be able to participate in the decision- making processes that shape our society. While we all had our heads down making a living, the world was taken over by actuaries, accountants, analysts, economists, and a host of others that use mathematics to determine everything from how much radiation is safe to your credit score. Mathematics is no longer limited to questions of gravity and atomic mass. It now plays an important role in every aspect of your personal and family well-being.

Today the questions and issues we face on a daily basis are more complex than in the past. Global warming, renewable energy, mortgage rates, funding for Social Security, and Medicare are just some of the issues that confront us on a daily basis. Understanding these issues and participating in the decision process requires greater knowledge of mathematics by more people. The gap between scientists and the rest of us will remain; however, the nature of the gap must be transformed. It is now necessary that everyone knows why mathematics is so important. Teachers must know, policymakers must know, textbook publishers must know, parents must know, and students must know.

Our education system has never been good at providing what every student needs in order to reach his or her fullest potential. Instead the system has only been asked to ensure that there are an adequate number of highly talented individuals that can be leaders in science and mathematics. The issue now is that we need the system to produce a greater number of these talented students and we need all students to learn more mathematics. This then will require a restructuring of the system to be more attentive to the needs of a larger number of students. A giant step forward will be telling everyone why we need to learn mathematics.

The Question

The question is asked because of the gap between the mathematics that mathematicians and scientists study and use and the mathematics that the rest of the world needs to know. Our current education system approaches mathematics instruction as if everyone were going to become a mathematician or scientist. However, we know that the system is not designed to make everyone a mathematician or scientist. Thus, those who have no intention of becoming a mathematician or a scientist are forced to endure instruction that is not responsive to their future goals and aspirations. Even the students who are interested in careers in mathematics and science may see little relevance between mathematics and the rest of their lives.

Three words in the question "Why do I have to know mathematics?" need to be made a bit more precise. We need to be clear on what we mean when we say "mathematics," "know," and "need." I have already attempted to clarify how the term *mathematics* is used in the question "Why do I have to learn mathematics?" The question does not apply to counting, adding, multiplying, and dividing. Most people understand a need for common fractions like halves, thirds, and fourths, eighths, and sixteenths. We use and see percents frequently enough to see utility in learning basic percents. Counting, multiplication, division, adding subtracting, simple fractions, and simple percents are what will be referred to here as basic mathematics or arithmetic, but when we leave the realm of basic mathematics and applications, students really begin to question the relevance of mathematics.

It is when students begin the study of algebra that they are most likely to begin to question the need to know mathematics. In this book, when I use the term *mathematics*, I am referring to the concepts, procedures, and applications commonly covered in algebra, trigonometry, calculus, differential equations courses, and even higher-level mathematics courses. By mathematics, we do not refer to the innate ability to judge distance, or to approximate the amount of time needed to complete a chore, or the mathematics that may be involved in any other number of activities that we routinely complete. These activities and abilities may play a role in our ability to do and learn more explicit mathematics, but they are not what we are considering here.

Another key word in the question is "know." What do we mean by the word "know" when we ask, why do I have to know mathematics? Are there different ways to know a thing? Are there different ways to know mathematics? Current measures of how much mathematics a person knows are based on how far he or she has progressed in the sequence of mathematics courses taught in high school and college. Despite many education reform efforts, assessment of knowledge in most mathematics courses is based on computational ability and or the ability to apply the correct procedure to compute an answer. Our education system is slow to let go of standard measures of knowledge and achievement. The question we address here and that is asked by millions of students everywhere reflects the inadequacy of this one way of knowing mathematics. The answer we will give will lead to a different way of knowing mathematics.

Imagine that you are formally introduced to someone with whom you have little in common. Suppose further that you are constantly thrown into situations where you must interact with this person. You can know mathematics this way. Now, suppose that you meet someone on the street and strike up a conversation. You may find that you have lots in common with this individual and that you share common interests. You may know mathematics this way. We don't usually form lifelong meaningful relationships with people or things that are forced upon us. We need to know mathematics up close and personal.

The last key word that must be clarified is "need." What do we mean by "need" to know mathematics? We need to breathe. We also need to exercise, but exercise is not as central a need as breathing. The more we exercise then the more we need to breathe. Mathematics is like breathing. We all need to count and add; thus some mathematics is clearly necessary. As has already been discussed, a number of people live highly productive and successful lives without possessing or applying any mathematics knowledge beyond the level of basics mathematics. The need to know mathematics is clearly relative to your career choice and the kinds of economic, social, and political activities in which you participate. The more you exercise, the more you need to breathe, and as we will see, the more you participate in the economic, social, and political decisions that shape your environment, then the more you will "need" mathematics.

Do We Really Need to Know Why Mathematics Is Important?

We have begun so far with a basic assumption that it is important for people to know mathematics and that it is important for them to know why they must learn mathematics. Some people might say that knowing why we need to know mathematics is not important. After all, we do not need to know why we have to breathe in order to breathe. Is it critical that we know why we need to learn mathematics? Perhaps we do not really need an answer to this question to benefit from the study of mathematics. It is quite possible to be a straight "A" student in mathematics while having little idea of why it is so important. So what is the big issue? Why not leave things the way they are? Do we have a shortage of engineers and other scientists? Some would say we do. Others say that we still produce as many scientists and engineers as we always have; however, they are choosing to pursue more lucrative careers in other areas. The apparent shortage of mathematics teachers is also a result of economics. Instead of deciding to become teachers, many good mathematics students opt for the better paying jobs they were promised for learning mathematics.

The problem is not that good mathematics students do not know why mathematics is important (and many do not). The problem is that not knowing why mathematics is important may help to cause the vast majority of students to lose interest in mathematics. The lack of meaning makes mathematics seem like no more than a series of definitions, rules, and procedures. The result is that many students build resentment and frustration toward mathematics and eventually stop studying it altogether.

These students have not failed to comprehend mathematics because of a lack of ability, but because of a lack of motivation, direction, and often poor instruction. Unfortunately, the end result can be that they do not get into the college that they desire or they may not get the job they want. If they do go on to college, they may struggle to meet their mathematics requirements, even though the reasons for these requirements in the major are not always clear. How many students change majors because of mathematics and choose a major that requires less mathematics—thus, reaffirming mathematics as the gatekeeper to better jobs?

Motivation is a product of the expectations for success and the perceived value of accomplishing the task. The fact that many students and parents indicate that their mathematics grades were great in elementary school only to drop in junior high and high school might suggest that at least students start with a good expectations for success, which leaves only the perceived value as the culprit to explain the drop in grades. Thus knowing why mathematics is important could have a significant impact on students' effort. Effort is based on the expectation of fulfillment of personal goals. So if students cannot see a relationship between mathematics and their personal goals, then it stands to reason that they would not put forth much effort to learn mathematics.

Studies have shown that while Asian students tend to score higher on mathematics achievement tests than American students they do not have greater personal motivation. Their success comes from a desire to please their parents who set high expectations and from a desire to fit in with the cultural norms that also set high expectations. Many American students do not have parents who set these high expectations in mathematics or do not try so hard to meet their parents' expectations, and of course, American culture sets a lower expectation for mathematics success. All of these are reasons why students need to know why mathematics is important.

Some Common Answers Explored

Okay, before we get to the actual answer to the central question, let's take a closer look at some of the typical answers that are given. We will see later that while these typical responses may fall short of providing the answers that students are looking for, they are in fact true.

We are told that you need to know algebra so that you can move on to the next course. While this particular answer does little to demonstrate the relevance of mathematics to anything other than learning more mathematics, it is a true statement. Algebra is to higher mathematics what grammar and syntax are to language. The concepts and procedures developed and learned in algebra are essential to success in higher levels of mathematics. In fact, algebraic skills and techniques are used at every level of mathematics. Algebra is the foundation on which higher levels of mathematics are built. However, this response does not explain why we need to go further in mathematics. In fact, the thought of studying even more mathematics may only add to the students' frustration. This response puts the question off to be asked again when the students begin to study calculus, if they make it that far. Those who do not go on to study calculus may never find an answer to the question.

Another popular answer is that you must learn mathematics in order to get into college or get a good job. This answer has much more potential to motivate students than the simple idea of studying more mathematics. Those who are motivated to go to college may be willing to accept this answer; however, even if they do, they still do not know how mathematics is relevant to their lives. They simply accept that the requirement must be met. This answer plays on the students' fear and may help to fuel the development of math anxiety. Many children who suffer from mathematics anxiety may associate a strong negative consequence with doing and failing mathematics tasks.

Once the students actually get into college, they may find themselves asking the same questions again. Only now they are already in college. So now why do they have to learn even more mathematics? Well, this is where the desire to get a job is used against them. They are told that if you want to get a good job and be competitive in today's economy, you have to learn mathematics. So just when they have gotten over the anxiety of getting into college, they are threatened with joblessness, and still they don't know how they will ever use mathematics in this grand competition for jobs and success. Good grades in mathematics can help a student get into the college of his or her choice, and learning mathematics can make someone more competitive in today's economy. However, since many college graduates do not see the value of their mathematics classes, the truth of these types responses is not clearly evident to many college students and does not provide the answer for which we are all searching.

Why Do I Have to Know Mathematics?

The number one and perhaps my favorite answer to the question why do I have to know mathematics is that mathematics is all around us. This one probably comes closest to providing the missing connection between mathematics and the students' social and physical realities. However, the statement itself grossly oversimplifies a number of important ideas. If mathematics is all around us, then why don't we see it? In fact, if mathematics were really all around us, then wouldn't the need to study it be obvious? While this answer opens the door for understanding why we must study mathematics, it falls short of putting the answer in our hands.

Okay, so is mathematics all around us or not? The answer is yes. Mathematics really is all around us, but it is not always in plain view. Some would suggest that when we estimate a quantity or a distance, or when we organize our time we use mathematics. While it is true that when we engage in these kinds of daily activities we use intuitive quantitative strategies, these activities do not justify the study of algebra, geometry, or calculus. Basic arithmetic is more than enough to get us through our day on most, if not all, occasions. But algebra, geometry, and calculus are around us everyday too. To see them you have to look in the right places. Later in the chapter, we will see where you have to look and why you should be looking.

All of these commonly heard responses to the question—why do I have to study mathematics?—are correct.

We do need to know mathematics for the next course, mathematics can help you get into college and can help you to get a good job, and it really is all around us. Yet given these answers, there is still something missing. These answers fall short of providing the relevance between mathematics and our daily lives that we all seek.

What Is Mathematics For?

We are almost ready to give the answer we have all been anticipating. However, to understand the reason you need to know mathematics, you must know that it is used for three things. Once you know these three things, then your need to know mathematics will be based on how much you engage in these three activities. Then the real question will be why are you not engaged in these three activities?

First, you must note that, with the possible exception of some physics and mathematical constants, everything changes. Nothing stays the same. Change is both natural and inevitable. Change is at the center of our existence, and without change, there is no life. Change is the reason that mathematics exists and the reason that it is all around us. Everything changes.

When something changes, the first thing we want and need to know is how much it changed. We want to measure the change. So the first thing that mathematics is for is, measuring change. How big was the change? How fast was the change? Measuring change can be a simple matter of subtracting one value from another, or it may be a more complex process involving percent change, relative change, or average rate of change. Since the first use of mathematics is to measure change, and since everything changes, then mathematics is truly applicable to everything, just as your teachers have always told you.

But wait—if everything changes and mathematics is used to measure change, then shouldn't the need for mathematics be obvious? You are right, and the need for mathematics would be clear if people actually measured change. While it is true that everything changes and mathematics is used to measure change, most people do not measure change. People do make simple measurements. They may subtract the amount paid from the balance due when making car payments. They may notice a change in gas prices or a change in the temperature, but most changes go unnoticed and unmeasured.

In the place that you are currently reading this book, everything around you is changing. Most of the changes that take place in your environment go unmeasured and unnoticed. Psychological studies show how frequently we fail to observe even sizable changes in our environment. Even the changes that we become aware of are probably not measured. There is no need to measure the changes in your current environment because they are small and do not cause alarm. Perhaps the fact that everything changes is the reason that we so frequently do not notice or measure change. We have become comfortable with the small changes that take place around us every day. The reason that using mathematics to measure change is not obvious is because we do not typically need to measure change in order to operate in our environments. We may notice a change in political attitudes, a change in the weather, or a change in the way we feel without making any measurements.

Change is such a normal part of our lives that it often passes like the time without being noticed. From time to time our attention is drawn to a particularly significant change, such as the sudden rise in the price of gas. At that point, we begin to look for reasons and explanations; however, since we do not have a series of measurements of gas prices and other related variables, we do not have the necessary elements to form a system to understand the change. There are market analysts who have these measurements and provide explanations, so we allow others to make the measurements and process the data to give us the results.

Once our attention is drawn to a change, the natural and obvious question is what caused the change? We are often too busy trying to deal with the change to actually spend the time trying to answer this question. This is where the gap between the amount of mathematics you need to know and the mathematics that mathematicians, scientists, and economic analysts use begins to form. It is, in fact, an activity gap. You do not engage in the activity that requires you to know and use mathematics, so you are not involved in explaining change.

The need to explain leads us to the second purpose of mathematics, which is to predict change. Predicting change is at the heart of developing explanations and understandings of our physical and social environments. If you have measured things like barometric pressure, humidity, temperature, wind speed, and precipitation over an extended period of time, then you can use this data to search for relationships between the changes in the various quantities. These relationships can be used to help you make predictions about the weather. The relationships we discover in the data represent one aspect of the knowledge that we have about how and why the system changes.

Predicting change is one of the most powerful things we can do in our environment. If we can predict change, then we can avoid change, we can plan for it, or we can prevent it. The ability to make accurate predictions allows us to alter our environments to conform to our specifications rather than accepting preexisting conditions and changes. However, if we do not measure change, then we cannot predict change and so we cannot understand it.

Often, instead of measuring change to make accurate predictions, we may rely on inductive reasoning based on past experiences. We use our past experiences to draw conclusions and make generalizations about our environment. We assume that what has happened in the past will continue to happen. Thus, we are able to predict traffic patterns for our daily commute and what we will have for dinner on Thursday night. However, we typically leave predicting the weather and the cost of gasoline to those that have been trained (through the advanced study of mathematics, meteorology, and business) to make those predictions. But how long can we or should we rely on others to tell us what and why things are changing? When we consider issues like global warming, even the weatherman's predictions need to be considered more closely and cautiously.

Once again, since everything changes, mathematics truly is all around us. We can use mathematics to measure change. Then we can use those measurements to look for and identify patterns and relationships that in turn allow us to make predictions. The reason so many people do not know why they have to study mathematics is because they do not measure change and so they do not produce precise predictions. We have meteorologists to predict the weather for us because our intuitions about the weather do not provide the accuracy that we need to plan our vacations, fishing trips, picnics, weddings, or test flights of secret new military aircraft. All of these things require bright sunny days, so we plan based on the predictions of the weatherman. It is because our activities require more precision and accuracy that we measure and predict. If change is the mother of mathematics, then precision is the father of mathematics, and things like physics, engineering, economics, and finance are the children.

The last thing that mathematics is for is to manage change. In some situations, I am able to control change, but often the best I can hope for is to manage it. If I can predict the weather, then I can manage it by planning my outside activities on days when the weather is nice. We depend on accurate measurements and predictions to understand and manage all of our systems. If you are measuring and predicting change, then mathematics gives you the power to manage systems in your physical and social environment for your best benefit.

In our world today, many important aspects of our social environment are managed and are manageable. Some systems are discovered in nature, and others are developed by mankind. We live in a world that is run by political systems, economic systems, healthcare systems, marketing systems, and distribution systems. All of these systems have impacts on our well-being, and all of them are being managed by others for their best interest. Who is watching out for your best interest?

The Answer?

So why do you have to learn mathematics? The answer is that you do not have to learn mathematics. What? That's right. You do not have to learn mathematics. You have read this far only to find out that you were right all along; you never will use mathematics. But that is the problem; if you do not need to know mathematics, then you are not engaged in activities that require measurement, prediction, or management of change. If you are not measuring, predicting, or managing change, then you are a victim of change. If you are not managing change, then change is managing you.

You may ask, "So what is the problem? I am getting along just fine without measuring, predicting, and managing change with all of the precision that mathematics makes possible." You may say, "I am handling change just fine." My response is that you are also breathing just fine, but the fact that you are breathing is not an indication of the quality of your life. Exercising and thus increasing the need to breathe could improve the length and quality of your life. Likewise managing the changes in your life will increase the amount of mathematics that you need to know, and it could also increase the length and quality of your life.

Mathematics was born out of the need to understand and explain nature and our physical world. Our need to understand nature has now been equaled by our need to understand the man-made systems that surround us. These man-made systems have both positive and negative effects on our well-being, and they have positive and negative impacts on our physical environments. Many people have learned to manage our political and economic systems to make millions of dollars for themselves. People who have healthy retirement portfolios are managing the system, or they are having it managed for them by someone with an even better retirement portfolio. Many are able to use the powers of mathematics to help them to manage systems in their physical and social environments, or they have hired someone to help.

People who are trapped in low-paying dead-end jobs are not managing the system to their best benefit. People who are struggling to pay bills or to pay for college for their kids are not managing the system to their best benefit. A lot of people are not able to use the power of mathematics to manage the system for their best benefit. In fact, there are so many people who do not manage the system that it seems normal not to manage it. We may even begin to think that we cannot manage the system. Managing change, like mathematics, is left to others, and thus the rewards for managing change, and so managing the system, are also left to others.

A common answer students give when asked why they must learn mathematics is that they must learn it so that they can count their money and not get cheated. In fact, a lot of people are counting your money. There are lots of people that use the power of mathematics to ensure that they take as much of your money as possible. Credit card companies use the power of mathematics to set interest rates and determine a payment schedule that ensures that they maximize their profits. That means they take as much of your money as possible.

Product manufactures do not simply access manufacturing cost of their product to set a sales price. They also use the power of mathematics to determine how much money you and I have and how much we are willing to pay for the product. What we are willing to pay may be significantly more than in cost to manufacture the product. They can charge more because marketing experts use the power of mathematics and psychology to increase our desire to purchase the product. The value of the product is determined by how much you and I are willing to pay. The point is that lots of people in the world use mathematics against you and for their best advantage. If you are not able to use mathematics, then you have brought the proverbial knife to a gunfight. You may survive, but you can't win.

If you are not able to use the power of mathematics to your advantage, then you cannot exert maximum control over your future. You cannot make sound predictions of future incomes, values, cost, or opportunities. You cannot assess the values of political proposals or sales pitches or determine when a company is taking advantage of your willingness to pay for inferior products and when the company is offering quality at a fair price. You are at the mercy of the people who can use the powers of mathematics to measure, predict, and manage change for their best interest.

Full participation in the decisions that face our nations and the world require greater understanding of change and so greater understanding of the mathematics that is used to measure, predict, and manage change. Now that you know why you have to know mathematics you need to ask a new question. How do I need to know mathematics?

If you are beginning to believe that you may, in fact, need to know more mathematics, then you may be wondering where to start. Do I need to take or retake algebra? Do I need to take a calculus or statistics course? The real and most important question is "What is the change in your physical or social environment that you want to understand?" You have to start with this question. Isaac Newton started with questions about how the solar system changes, which led him to the development of calculus. Statistics was developed from early efforts to keep track of births, deaths, and other civic data. Probability was first developed in the effort to understand and master games of chance. Instead of learning mathematics and then figuring out how and where to use it, the idea is to figure out what you want to understand first. What change or kind of change is it that you would like to understand? Your effort to understand change will drive and guide your need to know and understand mathematics.

In the chapters that follow, we will continue to develop our understanding of the role of mathematics in measuring, predicting, and managing change. We will also consider the important question of how you need to know mathematics. Do we need to know mathematics procedures? Do we need to understand mathematics concepts? Do we need algorithmic/computational skills? We will see that the answers to all of these questions depend on the kinds of change that you are trying to understand. What should be clear is that our goal is to be able to understand change.

Chapter Two

How Do I Have to Know Mathematics?

> Possession of an extensive vocabulary and exceptional knowledge of English grammar does not mean that you have anything of importance to say. Do great mathematicians have similar issues?

In chapter 1, we laid the foundation for understanding why you have to learn mathematics. In this chapter, we will discuss different ways that you can know mathematics and the ways that you will need to know mathematics to meet the expectations of our new knowledge-based economy. Then we will see a practical example of why you need to know mathematics. A key issue that will be emphasized throughout the book is that how you need to know mathematics will depend on the kinds of questions that you ask. There are some questions that we all ask (need to ask), so we all need to know some mathematics.

Given that we started the book by asking why you have to know mathematics, you may be a little concerned now to find out that there are many different ways to know mathematics. We will discuss three different ways to know mathematics. These different ways of knowing are more than just how well you know mathematics; they are three different ways of knowing. Each of the three ways of knowing mathematics is related to the other, and each can be the central feature in the way that mathematics is taught and learned in formal classroom settings.

Mathematics as a System: First, there is knowing mathematics as a structured system of concepts, definitions, formulas, and procedures. This way of knowing mathematics was the central feature of mathematics classrooms for most of the last 100 years. As you might expect, students in these classrooms would be the first to ask why they have to know mathematics. Teaching the basic concepts, structure, rules, definitions, and procedures in mathematics was typically approached in an abstract decontextual manner. This particular teaching approach helped to create the gap between mathematics and daily reality that leads to the questions concerning the relevance of mathematics. A certain amount of this kind of knowing is necessary to any of the other ways of knowing mathematics. However, in many cases learning concepts, structure, definitions, formulas, and procedures was, and still is, all that students learn about mathematics. This may be the way that many of you remember the mathematics classroom.

Teaching English can also be approached as a system of rules and definitions for communication. As with mathematics, there are certain rules and definitions that we must know to be able to communicate our feelings and thoughts. In English though, we quickly apply the rules and definitions to our daily realities, so the relevance of English instruction is less of an issue. English instruction is kept close to the activity level of the students. First graders do not write essays. As children get older, their communication needs grow more complex as does their English classroom instruction. The close relationship between classroom instruction in English and the students' lives is missing in the mathematics classroom that focuses on teaching abstract concepts, rules, and procedures. Students do not apply what they learn in their mathematics classrooms in their daily lives.

Even though there are a number of negative side effects, teaching definitions, formulas, and procedures in the abstract is an efficient way to teach and learn mathematics. There has always been a small subgroup of people who learn mathematics well when it is taught this way. However, this approach to knowing and teaching mathematics is not effective for the larger majority of students.

What's worst is that the majority was, and is, allowed to think that it was, and is, their fault that they didn't learn mathematics.

The emphasis on cost efficiency in the design of our education system makes it difficult if not impossible to ensure that all students learn high levels of mathematics. Instead it is enough to ensure that an adequate number of people learn mathematics at the level necessary to become scientists and engineers. The fact that the majority of students do not learn higher levels of mathematics is an expected outcome. The majority of people did not need to know higher levels of mathematics. Today the demands placed on citizens of the new global community dictate that more people learn more mathematics. This means that we must find ways to teach mathematics that are responsive to the needs of all students. This fact has been recognized by policymakers and mathematics educators, so there has been an effort to rethink what it is to teach and know mathematics. The results of these efforts lead us to the second way of knowing mathematics.

Mathematics as Problem Solving: The second way that we can think of knowing mathematics is using mathematics to solve problems. Problem solving is currently a major learning outcome for K–12 school systems across the country. Many teachers and curriculum developers stress the importance of students being able to use mathematics to solve problems. To accomplish this goal, courses are developed that present a series of contrived problems to be solved. As students move through the grade levels, the problems to be solved require different kinds of mathematics and more advanced mathematics. To know mathematics is to be able to apply general problem-solving strategies to solve real life problems.

This way of knowing mathematics gets closer to addressing the reason that mathematics is so important, but it misses the mark in an important way. It is difficult to develop a series of problems that will be real and or interesting to all students while maintaining a focus on a rigid mathematics curriculum. If the curriculum dictates that we cover polynomials, then we must have problems that involve polynomials. The problems that are developed in any such course are at best someone else's problem (not the students'), and at worst they are simply exercises designed to give students practice applying the target problem-solving strategy. In the first case, the student may find the problem presented in the text more or less interesting and/or relevant. In the second case, the whole course can degenerate into an irrelevant series of meaningless make-believe exercises that require students to compute answers to equations generated from contrived situations.

The problem-based approach to teaching has many good points but suffers from an unavoidable weakness. The problems are not real. In the problem-based approach to teaching and knowing mathematics, the problems are selected and designed to teach mathematics. But in the real world, problems arise for different reasons. Real problems do not come prestructured just so that a particular algebraic procedure is needed. When a real-world context such as gas prices or global warming is used in a problem-solving mathematics curriculum, the problems are predesigned to require a particular mathematics computation. In reality, understanding the real-world context is not the goal. This situation is made worse by the fact that the next problem in the text typically deals with a completely different context. Students quickly realize that these artificial problem-solving contexts are not important. The fortunate students are able to see that the real goal is learning the abstract mathematics concepts and procedures, but the unfortunate students are left with little more than a series of unrelated problem-solving experiences that have not helped them understand gas prices, global warming, or mathematics.

The problem-solving approach to teaching and knowing mathematics attempts to put a meaningful face on developing knowledge of mathematics as a structured system of concepts and procedures. It is an attempt to demonstrate and teach the utility of mathematical concepts and procedures through the solution of a series of problems. It is, in many respects, an attempt to make mathematics relevant. However the problem-solving approach fails to capture the true relevance of mathematics because, just like the mathematics, the problems have been abstracted from the context in which they arise.

A problem is an obstacle to achieving a desired goal. It is the objective that makes the problem relevant. Mathematics as problem solving places the focus on the obstacle; however, the obstacle is only relevant in terms of the broader context of the goal. Thus, mathematics as problem solving falls short of providing the relevance that many students are seeking when they ask why they have to learn mathematics.

Mathematics as a Way to Understand: If problem solving is the removal of obstacles, then what is the obstacle that mathematics removes? The answer to this question is, in fact, where the relevance of mathematics can be found. Since obstacles exist because of a goal, then a better question may be what is the objective that mathematics helps one to achieve. This brings us to the third and most important way of knowing mathematics.

The problem-solving approach to teaching and knowing mathematics is based on a false understanding of what mathematicians and scientists are doing when they are trying to solve a complex mathematical equation. They are not simply trying to find an answer to a computational algorithm; they are trying to understand a system or a process. For the mathematician and the scientist, the search for understanding is the driving force behind the development of systems of mathematical equations that represent and model reality. The goal that mathematics helps one to achieve is the development of understanding about change in our physical and social environments. Thus a mathematical problem exists when there is a gap between my current and my desired understanding of change.

The problem-solving approach to teaching mathematics does not capture the role of the quest for understanding that drives the development of mathematics. The third way of knowing mathematics is based on this quest for understanding. The third way that we can know mathematics is as a resource for understanding and more specifically as a resource for understanding change. We are prewired and equipped to recognize and respond to change. We also have a natural desire and need to understand change. In fact, mathematics developed in response to our efforts to understand changes in our physical environment.

You need to know mathematics as a resource for developing understanding. This is the one way of knowing mathematics that unites all other ways of knowing. It is the unification force of mathematics knowledge. All forms of knowing mathematics are derivatives of our quest to understand change in our physical and social environments.

The structure of mathematics is the structure of change, and learning about mathematical structure, concepts, and procedures provides the resources needed to organize and structure our efforts to understand change. Knowledge of the structure of mathematics and of mathematics as problem solving provides the skills needed to develop a quantitative conceptualization of a context and then to use this conceptualization to compute meaningful answers. Thus we must know about the structure of mathematics, and we need to know mathematics as problem solving, but we must apply this knowledge toward the development of understanding if we are to know and realize the true purpose of mathematics.

Mathematics concepts are used to help us organize and understand change, and it is our interest in and drive to understand change that focus our need to know mathematics. When the goal of mathematics instruction is to develop our ability to understand, then there is no need to ask why we need to know mathematics since we are using it to accomplish a relevant task. The key issue in teaching and knowing mathematics as a way to understand is that we must ensure that students want to understand something; rather we must ensure that we do not remove the desire to understand from the student.

Maintaining a Desire to Know: Mathematics provides the resources that we need to organize and understand change in our physical and social environments. It is our drive and interest in change that fuel our need to know mathematics. How the desire to understand may be removed or sustained by schools is an important issue that deserves much more investigation and explanation. However, while a full treatment of this issue would take us too far from our current goal, a little explanation is important. We are, in fact, wired from birth with the ability to recognize change and a desire to understand and explain it. However, by the time we get out of high school, many of us have lost or suppressed these desires. How does this happen, and what role does formal schooling play?

When you arrived at the first grade, you were like all children, curious and full of questions. One of the things that your first grade teacher was required to do was to teach you how to do school. Thus, when you asked your teacher, "How do door knobs work?" or "Where does wind come from?" and "Why is the sky blue?" you were told that you had to sit down and raise your hand when you wanted to ask a question. So you rushed to your seat and raised your hand. When the teacher called on you, you asked, "Why is the sky blue?" The teacher's reply was that you must learn to focus on what the teacher is discussing.

By the fourth grade, you had learned the rules of school. Now you sat and listened attentively to what the teacher was discussing, and you raised your hand to ask questions about the topic the teacher discussed. You no longer asked questions about the things that you found interesting because you knew that that was not why you went to school.

As you grew a little older, your views of the world began to change, and you were less impressionable. You had a lot more questions than you did when you were six. However, now you knew that you could not ask those questions in school. The result was that you stopped participating and you lost interest in what the teachers were discussing. Finally, when you reached high school, you began to see the light at the end of the tunnel, and you had only one question: "Is it going to be on the test?"

This scenario may more or less accurately reflect your experiences in school. If you did not manage to leave high school with a strong desire to know and understand, then there is a good chance that you are not a scientist or engineer. In the process of making schools and classrooms manageable environments for teaching and learning, we can inadvertently cause students to lose or suppress their natural desire to know. This is counterproductive to the purpose of school; if you do not want to know anything, then you certainly do not need to know mathematics, science, or even how to read beyond reading a lunch menu and the TV listings. Thus, one of the most important things we must do to support learning of all kinds and support mathematics learning in particular is to ensure that students retain their curiosity and desire to know and understand.

Mathematics for Understanding/Quantitative Literacy: Let's return to our main topic. We know that mathematics is for measuring, predicting, and managing change and that we measure, predict, and manage change as a means to develop understanding. We must rethink what mathematics we need to know and how we need to know it. As was stated, the use of mathematics to develop understanding is a unification force. It does not replace the other two ways of knowing. Mathematics as a way to understand gives purpose to knowing the structure of mathematics concepts and procedures and a reason for developing our problem-solving abilities. We must still learn some definitions, formulas, and procedures, and we still need to develop our problem-solving ability (i.e., the ability to apply standard mathematics procedures toward the computation of answers to equations that arise from the analysis of situations in our physical or social worlds). The question now is not why do I have to know mathematics, but what mathematics do I have to know and how do I have to know it?

You may see the terms *numeracy*, *matheracy*, and *quantitative literacy* being used to describe new mathematics knowledge requirements needed to participate in our new global knowledge economy. There is some debate over the definition of these terms. It is not even clear that they all refer to the same thing. We will use the term *quantitative literacy* to represent our new mathematics learning goals.

In part, the debate over the use of one term or another and debates over the meaning of the terms is caused by conflicting and vague understandings of what mathematics is required for quantitative literacy and why quantitative literacy is even needed. Vague suggestions that quantitative literacy is needed to deal with the quantitative aspects of adult life or the importance of being able to handle the practical mathematics demands of everyday life do little to help clarify the situation. As we already know, most people do not know mathematics beyond the basics, and yet they are apparently able to handle the quantitative aspects of adult life and the practical demands of everyday life. That is unless the practical demands of everyday life have changed.

It is this last point that is the key. The practical demands of everyday life have, in fact, changed. In the past, it was not practical for the average citizen to understand global markets, national healthcare, or global environmental concerns. Today our world has become a much more complex environment. In today's world, it is important that the average citizen be able to use mathematics to develop the understandings needed to participate in social decisions and local, national, and global governance.

Focusing on the use of mathematics to measure, predict, and manage change to develop greater understanding provides the clarity needed to define quantitative literacy. With measuring, predicting, and managing change to develop understanding as our starting point, it is easier to define what mathematics is needed and how well it needs to be known. A focus on measuring, predicting, and managing change situates mathematics teaching and learning in real contexts where it is used to accomplish real goals and objectives.

There are, however, still a number of basic mathematics concepts and procedures that we will have to learn. This is not unlike learning to read, write, and speak a language. We must learn an alphabet, rules for spelling, grammar, syntax, and of course vocabulary. In mathematics, we must learn numbers and symbols, operations, properties, and rules for operations. We then use our basic understanding in language and mathematics to develop more advanced concepts and skills as we apply the basic understandings to our efforts to describe and understand our surroundings.

If we have a good understanding of the basics, then it is easier to build new knowledge. A weak foundation in grammar, syntax, and spelling will limit my ability to produce and consume written texts. Similarly, a weak foundation in counting, addition, multiplication, fractions, decimals, and percents will limit my ability to produce and consume mathematical descriptions of change. The good news is that students do not generally question the need to learn these basic mathematics concepts. The bad news is that we have not always done a good job teaching these concepts. Part of this is because we do not give students the opportunity to use what we teach as we do in English classrooms.

While some may disagree, there is a great deal of memorization that needs to take place in the early part of mathematics learning. There are also a number of basic skills that need to be developed. The same is true in English; however, when students learn vocabulary and parts of speech, they put this knowledge to use right away in the effort to share "their own" ideas with others. However, in mathematics when students learn the basics of counting, adding, division, percents, and fractions, they are not allowed or encouraged to use this knew knowledge in "their own" quest to understand or communicate their ideas about how something is changing, how it will change, or how it has changed.

An important aspect of quantitative literacy is the ability to communicate one's own ideas about change and to understand the ideas communicated by others about change. In this regard, quantitative literacy is not unlike any other form of literacy. The goal is clear communication. The difference lies in the fact that what is being communicated is one's understanding of how something is changing, has changed, or will change. The mathematics needed beyond the basics of counting, adding, division, percents, and fractions will depend on what kind of change I am trying to understand and/or describe.

Beyond the Basics: When we ask what mathematics is needed beyond the basics, we might just as well also ask what words we should teach for literacy. There are, of course, basic common words that all should know, just as there are basic and common ways to discuss our understanding of change. It would be difficult to make a list of words or mathematical concepts that everyone would agree to as the necessary list. There are advanced words, syntax, and grammar that are used to improve communication, and there are advanced mathematical concepts and modeling techniques that improve the ability to understand and describe change. In language, we shift concern from a specific word list to the students' ability to convey meaning and to relate their experiences. Using mathematics to measure, predict, manage, and understand change to drive the curriculum shifts the focus from what mathematics to teach to what kinds of change students should be able to describe and understand. How well can or should a student be able to communicate their understandings about change?

This might suggest a different order of topics as well as the elimination of some topics. It will also require that more focus be placed on the students' ability to learn mathematics content independently. We must think in terms of building capacity to learn. Once I begin to use a basic vocabulary, then I am better prepared to learn more vocabulary. If I have experience playing football, golf, or tennis, then I have a greater capacity for learning the language associated with these sports. If I have experiences measuring, predicting, and managing change, then I have a greater capacity to learn the mathematics associated with these activities.

Technology: The use of technology will greatly change the ways that you will need to know mathematics. This is in large part because you will be relying on technology to do much of the calculation. Measuring, predicting, and managing change can often require labor intensive and tedious computations that are highly susceptible to errors caused by oversight and mental mistakes. Technology aids in these computations and also frees the user to focus more attention on the outcomes than on the computation itself. Computation has always been a necessary evil and an end in itself. The fact that technology can relieve us of much of that burden should be embraced.

There are mathematics purists among us who will argue that using technology to do computation is bad because it reduces our understanding of the subject. For example, the use of calculators in elementary schools has both positive and negative consequences. I am appalled to see young students reach for a calculator to perform even the simplest additional and multiplication tasks. I am sure that their weakness in this area must limit their ability to brainstorm when solving problems. However, at the other end of the spectrum, computers are indispensable tools used by mathematicians to perform tasks that are impossible to carryout by hand.

This will not be the first shift in the way that people know mathematics. Ancient mathematicians and scientists would be appalled to see the wanton lack of logic taught in modern mathematics classrooms. Over the past thirty years, the role that proof plays in mathematics instruction has been reduced in favor of a greater focus on applications. We will have to accept that using technology will change our understanding of mathematics. It is not clear whether the change will be a reduction in a general sense. It may simply lead to a different way of knowing. There will still be a small group of students who study mathematics in the most traditional sense. The potential change in the way we know mathematics is a great example of a change that will have to be measured. A bigger and more important change is the fact that people will actually be using mathematics.

Two Different Relationships with Mathematics

There is a difference in the relationship between mathematics and the physical world and between mathematics and social world. We typically fit mathematics to reality. We make observations of changes in our physical surroundings, and then we use mathematics to create models that represent the relationships between the changes in one component and another. The validity of the model depends on the degree to which the model is able to predict events in the physical world. For example, we have models that represent the speed of falling objects. If you drop a ball from a given height, we can use the mathematics model to determine how long it will take for the object to hit the ground. The model is very accurate. Changing the mathematics does not change the way the object falls.

We also have models that predict the weather. We are all familiar with the accuracy (or the lack of accuracy) of weather prediction. The weather is a much more complex system than a simple falling object; however, the validity of the model is still based on its ability to predict events in physical reality. Changing the model does not change the weather. The world is not obliged to do what our equations say it must. Instead, we must alter our equations to match what the world shows us.

Mathematics does not have a will to impose on the physical world; it can only represent what happens so that we can improve our understanding. Once we understand a physical system like falling objects or the weather, we can then apply our free will to make an impact on the physical systems, sometimes for good and sometimes not for good. We can make object fly, and we can recognize the negative impacts of pollution on our climate.

The fact that we are able to model and understand physical phenomenon like gravity, the solar system, light, sound, and energy using mathematics leads some to believe that God must have used mathematics when creating the universe. While we may never know if or how God used mathematics, we do know how man uses mathematics. Politics, government and private institutions, business, and economics were all developed with the help of mathematics. Mathematics plays a significant role in the development of these man-made systems. In our social environment, the results of mathematical models are not only used to check for consistency between models and reality, but in the social world, the mathematical results are used to change social reality. In our social reality, mathematics does impose its will in significant and important ways. We must do what the numbers tell use to do and change our system to match what the mathematical models tell us.

This makes mathematics even more important in social systems than in physical systems. Physical systems will continue to do what they do regardless of our attempts to understand, but social systems only work because of our ability to understand them. Those who understand know how to make the system work for their best interests, but those who do not understand cannot manage the system for their best benefit. Thus the idea that physics, chemistry, and engineering majors are the only ones that need advanced study in mathematics is incorrect. All students need to understand how to use mathematics to understand how economic systems, business systems, political systems, insurance systems, healthcare systems, and other systems change.

Perhaps it all started with the establishment of organized government that allocated resources and services and collected taxes to pay for government services. Everything was based on numbers. How many people? How much grain? How much to tax? What service to provide? Our system today has grown more and more complex, and the complexity is all supported by numbers and equations. Governments and businesses collect data to measure every facet of our lives, including leisure time activities, work habits, education levels, religious affiliation, spending patterns, political beliefs, and much more. The data is collected and analyzed, and the results are used to set prices, set benefit levels, set tax amounts, and allocate resources like police and community libraries.

We all assume that the process is conducted by individuals with the highest levels of moral and ethical fiber, and that they apply a totally objective and impartial eye to the analysis process. What if we are wrong? What if the people collecting an analyzing the data are not as honest, objective, and/or impartial as we believe? Suppose that their motives are not always in our best interest. Maybe we had better check up on them.

Later we will consider health care, a very important man-made system. Insurance companies, pharmaceutical companies, hospitals, and the government use mathematics to determine service requirements, production levels, coverage cost, and more. However, the key issue is that in the social world man is able to impose his will on the process and use mathematics to determine how to make his own preselected outcomes a reality. In other words, man uses mathematics to construct the social system the way he wants it to be.

We can also use mathematics to determine what should happen if we remove man's free will. What would happen if the system were allowed to continue on its own without new rules and regulations? We can use mathematics to consider different possibilities. In the physical world, we can imagine a world with less gravity and create mathematical models to represent life on such a planet, but we cannot create it. In the social world, we can imagine a health care system that does not cost 2.2 trillion dollars, and we can create it. But only the people who would benefit from such a system will work to create it. The problem is that those people are not quantitatively literate. I am talking about the masses.

A primary concern in discussions about quantitative literacy is the role that mathematics plays in good citizenship. There are a number of problems, issues, and systems that citizens of the new twenty-first century need to understand. Perhaps none is more important than local, national, and international economics. One of the most important ways that you need to use mathematics is to understand economics so that you can fully participate in and manage your own economic environment. It is the attempt to understand issues like economics that gives meaning to mathematical concepts. When mathematics is reconnected with understanding, then mathematical concepts take on new meaning and clarity.

When we try to understand how and why something changes, we first break the process into observable parts, pieces, or components. We then attempt to identify cause and effect relationships between the parts of the process that we have identified. We may see that a change in one part or variable causes a corresponding and predictable change in another part or variable. For example, when gas prices go up, spending in many other areas go down. It is the economic relationships between gas prices, travel, retail sales, and many other economic variables that give meaning to the mathematical symbols and expressions like factors, polynomials, linear, and nonlinear relationships that you see and discuss in mathematics classrooms. The reason people do not see the meaning of mathematics is that they are not using mathematics to make meaning of their physical and social realities.

An Example of Quantitative Literacy: Understanding Health Care

Life has meaning that is expressed by mathematical formulas. The search for understanding in life is what leads to understanding of and with mathematics.

Health care is a major economic system/variable that all good citizens need to understand. Health care costs are rising at an alarming rate. In 2007 as a nation, the United

States spent 2.2 trillion dollars on health care. The excessively large price tag impacts every area of economic activity. Employers struggle to provide health care benefits. The more employers have to spend on health care benefits, the less money there is to run the company and create more and better-paying jobs. Medicare and Medicaid consume more and more of federal and state budgets, which causes cutbacks in other areas and the laying off of state employees. Millions of people struggle to pay health care costs that are not covered by their insurance, if they are fortunate enough to have insurance.

I, like many other average citizens, would like to know why medical costs are so high. To understand and make sense of health care, we must identify the major factors and look for possible relationships that provide some insights in the causes that are driving health care costs up.

The cost of health care is dependent on a number of factors. Doctors, patients, pharmaceutical companies, lawyers, politicians, government officials, and insurance companies are some of the key players in health care. The general health of the people in the country is a factor, and I am sure that there are many more important factors. I do not claim that my analysis meets the standards of a professional financial analyst or an economist. Instead I am just an average citizen attempting to understand a complex issue. I am sure that my efforts will include simplifications of some of the complexity in the health care system;

however, these simplifications are necessary to begin the process of understanding.

As I began my investigation of health care, I asked myself what a good citizen should do and know relative to health care in order to make informed decisions. The process that I undertook was both instructive and revealing. I started by reading a few articles on health care. As I read articles from many sources, in many cases I was struck by the lack of objectivity in the articles. It was easy to tell if the author of the article was for or against health care reform. One author would give data to argue that insurance company profits were a major reason that health care costs have risen so fast, while another would present data to show that insurance company profits have not changed dramatically in recent years. It became clear that the ability to be critical of information and arguments provided by a growing number of reporting sources is one reason why we all need to develop our quantitative literacy.

The articles that I found on health care provided finished data analysis that was used to develop biased arguments to support or oppose health care spending reforms. I found myself desiring to see the raw data so that I could draw my own conclusions. I wanted to see for myself. The desire to know was critical to continuing my investigation and is critical to the development of quantitative literacy. To be quantitatively literate, you must be in the habit of questioning and critically analyzing information and data.

All the while that I was investigating healthcare reform, I was asking myself what a good citizen should be willing and required to do. Note that one answer is that we should be willing to read articles that present arguments for and against an issue; however, as a citizen, I should be able to expect that both kinds of articles will provide an objective analysis of the data. These articles should also provide evidence to support the analysis as well as supplying the data source. Perhaps one of the side effects of quantitative literacy is that people will begin to demand greater access to data and more responsible reporting. Again this is an important part of good citizenship.

Perhaps the most powerful role that quantitative literacy has to play in society is to help the average citizen to be more critically involved in the economic and political decisions that shape our world.

Understanding Health Care

In 1960, we spent 27 billion dollars on health care in the United States. In 2007, health care expenditures in the United States were more that 2.2 trillion dollars. Did you know that if you lived to be 100 years old and if you spent $1,000,000 every hour from the time of your birth you could not spend a trillion dollars? In 2007 there were approximately 302,000,000 people in the United States. We spent more than $7,400 dollars per person on health care. These numbers are so large that they become incomprehensible. Why does health care cost so much money?

We can blame insurance company profits, pharmaceutical company profits, or doctors' salaries. But while these factors are certainly a part of the problem, they do not by themselves account for the astronomical rise in health care cost. To truly understand the problem, we need data. The good news is that the government provides a lot of data. The bad news is that the data is not always easy to get, even when you know where it is.

Based on the articles that I read and the information from news broadcasts, I developed a list of the major variables that seem to play a role in health care cost. Here I must appeal to you to remember that when it comes to my understanding of the issues of health care and the economy I am an average citizen and not an economist. Thus my analysis is very much like what an average citizen may be expected to perform and does not represent or pretend to be any more than that. On the other hand, I would hope that those who do this kind of analysis professionally might see the need to provide better and more informative analysis geared to the average quantitatively literate citizen.

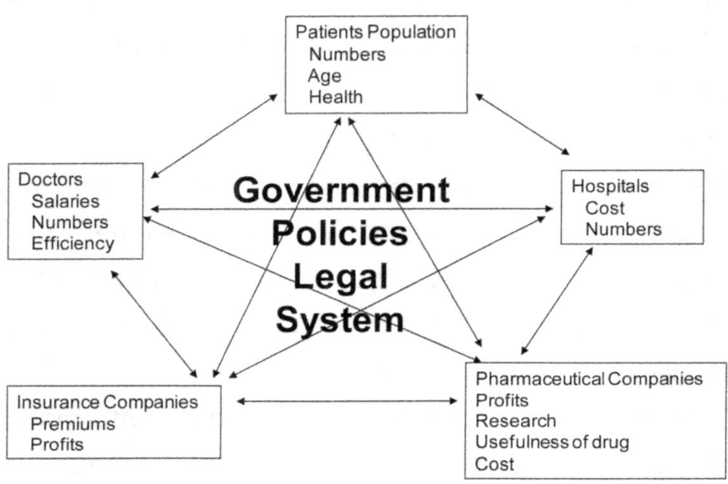

The model represents the highly complex interactions between some of the large number of variables in our health care system. You don't need mathematics to know that there is a relationship between the variables, but you need mathematics to understand the relationship. I know that rising hospital costs lead to rising health care costs, but what percentage of the increase is due to hospital costs? Are hospital costs rising faster or slower than overall health care spending? To understand the relationship, we must measure changes so that we can look for patterns and try to discover a relationship between the change in one variable and the change in another.

After developing my model as a guide I wanted to find data that would help me to answer a few basics questions.

I began by trying to see if costs were being driven up from the bottom by pharmaceutical and insurance company profits. How have insurance company profits on health care changed over the past 20 years. How have pharmaceutical companies' profits changed over the past twenty years? Unfortunately the data to answer these questions is buried in financial reports filed with the Security Exchange Commission. This data is not easily accessible to the average citizen although it should be. Although the limited data available suggest that pharmaceutical companies and insurance companies are a part of the problem, they do not seem to be the major source of rising costs. My analysis here was cut short because of a lack of access to the data.

I decided to look some place else to find a reason for high health care cost. I turned my focus on the people who get service from the system. Patients are at the top of the model because the medical system exists to provide service to the patients. Patients are the consumers of health care, so the cost should be driven by the number of patients and or other variables like age and overall health of the population. With this idea in mind, I set off to find data to help investigate the relationship between changes in the population and changes in health care spending.

After several false leads, a search for data on the Internet eventually led me to the Census Bureau, the Center for Disease Control, and related sites that provided a great deal of data collected from regional and national surveys on a wide range of topics. The government is to be commended for making the data available and for providing tools for accessing the data. However, finding and accessing the data that you want is not necessarily an ordinary citizen activity. There are some useful tables and charts, but much of the data is in databases that require special software to access and to process. Perhaps another of the outcomes of increased quantitative literacy in the population will be an increased demand for easy access to useful data.

I continued the search for understanding by considering the national health care expenditure data provided by the Census Bureau. Some of the data from this chart is provided in Table 1. You can review all of the data at the www.census.gov. As I said, it seems to be a reasonable assumption that the significant increase in spending on health care is due to increases in the population of the United States. If health care costs are being driven by increased demand for service (as many have argued), then we should be able to find a relationship between the number of people in the country and the amount of money spent on health care. A rise in the population of a certain size should lead to a predictable rise in health care cost. However, a comparison of the data shows that no such reasonable relationship exists. In fact, the comparison leads to alarming results.

NATIONAL HEALTH EXPENDITURES BY TYPE OF SERVICE AND SOURCE OF FUNDS: CALENDAR YEARS

LEVELS in $Millions	2007	2006	2005	2004	2003	2002	2001
National Health Expenditures	2,241,208	2,112,668	1,980,603	1,854,840	1,734,932	1,602,284	1,469,359
POPULATION	302	299	296	294	291	288	285
Health Svcs and Supplies	2,098,107	1,976,104	1,850,362	1,733,109	1,623,141	1,498,289	1,375,908
Personal Health Care	1,878,275	1,765,498	1,655,149	1,550,239	1,447,507	1,340,310	1,238,309
Hospital Care	696,539	649,327	607,497	566,828	527,380	488,369	451,238
Physician and Clinical Services	478,768	449,697	422,240	393,644	366,745	337,893	313,175
Dental Services	95,171	90,487	86,391	81,476	76,862	73,341	67,523
Other Professional Services	62,001	58,697	56,014	52,866	49,009	45,585	42,751
Home Health Care	59,031	53,038	48,085	42,708	38,025	34,213	32,179
Other Non-Durable Medical Products	37,357	35,342	33,991	32,720	32,067	30,405	29,951
Other Per'l Hlth Care	66,167	62,472	56,874	53,252	50,378	46,349	41,915
Admin. & Net Cost of Priv. Hlth Insurance	155,739	150,356	138,655	128,843	121,907	105,842	90,640
Public Health Activity	64,093	60,250	56,559	54,027	53,727	52,137	46,959

Table 1.

We can begin by comparing the percent change in the population to the percent change in health care spending. We use percent change in some cases because the actual difference is not as meaningful. To say that the population grew by three million people is more meaningful when we know how large the population was at the start. If the starting population was five million, then we get a sense of a dramatic change in population. However if the starting population was 300 million, then the change is still significant but not so dramatic. The difference is in the fact that three million is a larger percentage of five million than of 300 million, so the change feels bigger.

Percent change is also a useful way to compare changes in things that may not otherwise be comparable. If two variables are measured on significantly different scales, it may be hard to compare a change in one to a change in the other. If a grown man who weighs 170 pounds gains seventeen pounds, then he would experience a 10% weight change. A premature infant who weighs four pounds and seven ounces would need to gain about seven ounces to experience the same percent weight gain. In our typical daily experience, 7 oz is an incomprehensible amount. Comparing the two weight gains as a percent change helps to make the magnitude of the infant's weight gain more comprehensible. However, the comparison is only useful in one direction. The infant's weight gain offers little in terms of clarifying the adult's weight gain experiences. Also, a 10% weight gain for an infant is a good thing; a 10% weight gain for most adults is not. *(Note: This conversion is much easier in the metric system. In fact, a convert pounds and ounces to grams in order to compute the percentages. Perhaps another outcome of increase quantitative literacy will be that the United States will finally convert to the metric system.)*

Looking at the data we see that a population change of 1% corresponds to a 10% increase in health care cost. Increases in the population may contribute to rising health care costs, but there is no reason that a 1% increase in the population should lead to a 10% increase in health care spending. In the strictest case, we could say that there is a relationship between healthcare spending and population growth. Health care grows ten times faster than the population. This is an alarming result. In reality, it seems clear to me that population growth is not the reason health care is growing at such an alarming rate. A 1% increase in the population does not require or justify a 10% increase in healthcare. This finding alone, using a minimal amount of quantitative literacy, leads me to wonder how people could suggest that we should not do anything with regard to health care spending. Insurance coverage is not the issue. The cost of service is where the real problem lies.

We can make this more meaningful by relating it to our previous example. Let's compare the percent changes in health care and population growth to our adult male and our infant. If we do, then we see that the population is growing like a typical adult male, gaining about 1% of his body weight per year, while health care spending is growing like we would like to see a premature infant grow. We all know how fast babies grow anyway and how quickly they outgrow their clothes. Healthcare is quickly outgrowing our ability to pay, but like the infant, we are reluctant to put health care on a diet for fear of sleepless nights. At this rate, the baby will eat us out of house and home in a few more years.

A Closer Look at Healthcare Cost and Population Growth

So far, the level of quantitative literacy involved in the analysis is a comfort level working with numbers, percent change, and the ability to place numbers in a meaningful context. The rest is a matter of reasoning about health care. In fact, reasoning about health care or another important issue plays the biggest role in the investigation process and in the development of quantitative literacy. Let's use a little reasoning and some basic mathematics to look a little closer at the relationship between population and health care spending.

If we divide the amount spent on health care in any given year by the population, then we get the amount spent per individual (per capita) for health care. If we multiply the yearly increases in the US population by the amount of money spent per capita for health care, and add that to the health care expenditures for the previous year, then we should get the increase in health care expenditures as a result of population. In 1960, the United States spent 27.53 billion dollars on health care. There were 186 million people in the country. If we divide 27.53 billion dollars by 186 million people, we see that in 1960 we spent $148 per person (per capita). In 1960, the population increased by 3 million people. If we multiply the amount of money spent on each person by the 3 million new people, we get the expected increase in health care spending for 1961. Based on this calculation, health care cost in 1961 should have been 27.97 billion dollars (an increase of .4440 billion dollars). The actual healthcare expenditures for 1961 were 29.37 billion dollars. This is an increase of 1.83 billion dollars. Our calculation predicted an increase of .4440 billion dollars. Thus, 1.386 billion dollars of this increase was not due to population growth. The actual increase is more than four times larger than the increase should have been if the rise in cost was due to population alone. This is an unexplained increase in your health care bill, and you should demand an explanation.

In order to compute the expected increase in healthcare spending for 1962, we could use our number (i.e., 27.97 billion) or we could use the actual healthcare expense for 1962. By using the number reported by the government for 1962, we can allow for inflation while also allowing for the large as yet unexplained increase in cost. In 1962, we spent 155 dollars per capita on health care. The population rose again by about three million people, so the increase in health spending due to population growth should have been .4650 billion dollars. In fact, the increase in health care spending from 1961 to 1962 was 2.683 billion dollars. This is almost six times bigger than the increase based on population alone.

If we jump ahead, we see that in 2006 we spent $7,065 per person on healthcare. The population increase was again three million people. If we multiply these numbers, we see that there should be an increase in health care spending due to population growth of 21.2 billion dollars. So in 2007, we should have spent 2.133 trillion dollars on health care. In actuality, we spent 2.241 trillion dollars. This is an increase of 129.5 billion dollars. The increase is more than six times what the increase should have been if the rise in cost was due to population alone. The unexplained increase has grown from a little more than one billion dollars in 1960 to a little more than 100 billion dollars in 47 years. And now we really need an explanation.

Health care expenditure in billions from 1960 to 2007

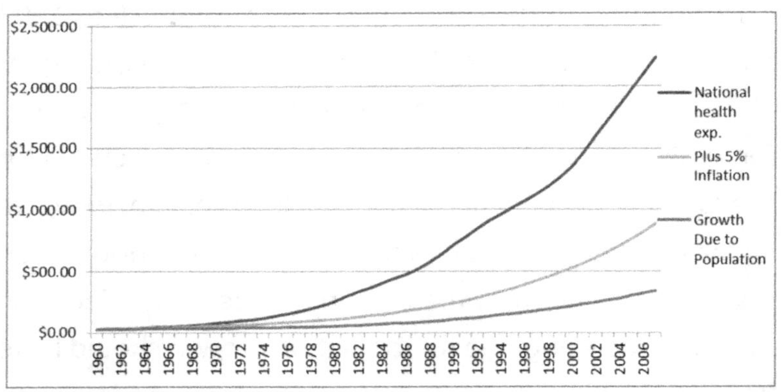

Chart 1:

In chart 1, the top line represents the actual health care expenditures, while the bottom line represents the growth in health care that would have resulted as a function of population growth. Based on the population growth model, we should be spending 340.35 billion dollars on health care in 2007 while in actuality we spent 2241.20 billion dollars, or 6.59 times as much as the size of the population would predict. Perhaps we did not allow for enough inflation in our computation. The middle line includes a 5% yearly inflation factor. Even assuming a 5% inflation level (much higher than the actual inflation rate), we should only be spending 884.38 billion dollars on health care. I am sure that the calculations used to develop these graphs do not account for a number of issues and variables, but that is exactly the point. The calculation shows that population growth is not the factor that is driving up health care costs. So what is it that we have not accounted for that is causing the sharp rise in cost? Whatever it is, it cost 1.35 trillion dollars a year, and it's growing.

I don't want the main point of this discussion to get lost in the investigation. That main point is that you need to ask quantitative questions. I am substituting my questions and analysis here only as an example. I am not trying to make an argument one way or another concerning health care. I am asking questions in order to improve our understanding. My next thought in that vein was that perhaps the rise in cost was associated with the aging of the population. As the population aged, there would be more people on Medicare. However, an analysis of this data, by again comparing percent changes, suggests that there is no predictable relationship between growth in Medicare enrollment and growth in Medicare spending. The data for this and other analysis can be found at www.cms.gov and www.census.gov.

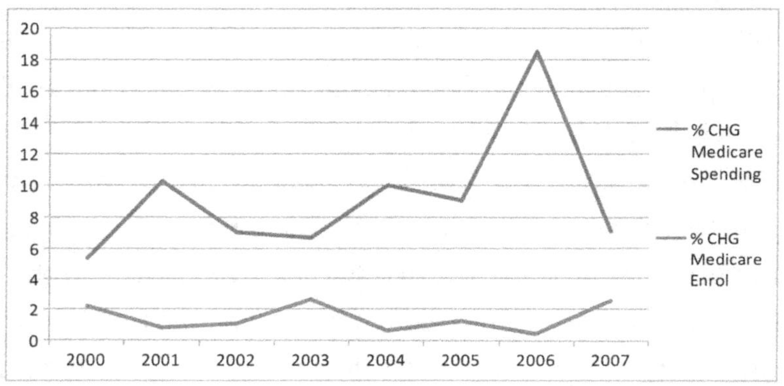

Figure 2

We can see from the graph in figure 2 and from the data in table 2 that there is no relationship between the percent change in enrollment in Medicare and the percent change in Medicare spending. The percent change in enrollment in 2001 was less than 1% while the percent change in spending was more than 10%. When the percent change in enrollment is small, the percent change in spending is large, and when the percent change in enrollment increases, the percent change in spending actually decreases. This is again a confusing result that requires an explanation. We debate about the number of people with and without medical insurance, about public options, and access to Medicare, yet no one is offering an explanation as to why the costs are as high as they are. In 1960, we dreamed of being a millionaire. Now we dream of being billionaires, and 1.5 trillion dollars a year can produce a lot of billionaires and millionaires. Has the number of rich people in the country grown as fast as health care spending?

Medicare spending and Medicare enrollment 2000 to 2007

2000	2.212712943	5.425890336
2001	0.802861685	10.25004457
2002	1.064584812	7.065092154
2003	2.621722846	6.753690061
2004	0.626013788	10.06900845
2005	1.193864443	9.052785151
2006	0.413171715	18.54654538
2007	2.558064596	7.19097381

Table 2

In the health care system, everything has gone up. It is like a bucket full of Styrofoam, wood, and plastic balls. The fact that all of the balls are rising cannot be blamed on the wooden balls more than the plastic balls. Just because the Styrofoam balls are on top does not mean that they are the cause for all of the balls to rise. In reality, the balls are rising because more water is being added to the bucket.

All of the balls rise. The debate is not over which ball is rising the fastest; we have to find the source of the water.

Cars cost more because people want them and are willing to pay for them. If a manufacturer makes a car that no one wants, then the car will not sell. If you are willing to pay $20,000 for a car, why would the dealer sell it for $15,000? Car prices are set by supply and demand. The price of medical treatment is not set by supply and demand.

The medical industry makes things that people want and sometimes need. How is the price set? Who would argue about the cost of a heart value to save his or her life? However, suppose that you had to pay for the heart valve yourself. When is saving your life worth the possibility of financially ruining the lives of your children? But we do not have to pay for the heart valve by ourselves. Instead private insurance companies, Blue Cross, Blue Shield, Medicare, and Medicaid spread the cost of the heart valve across the population—in effect, slowly financially ruining all of our lives (not just your kids' lives). The people who pay the bill set the price (by their ability and willingness to pay). Since we don't have to pay for the valve directly, we do not set the price of it. But then, who does set the price?

Perhaps an example of a study conducted by Ran Barniv, Kreag Danvers, and Joanne Healy in 2000 will help to address this last question. This study was randomly selected from many possible studies based on the title "The Impact of Medicare Capital Prospective Payment

Regulation on Hospital Capital Expenditures." It seemed that such an article might be relevant to my attempt to understand why health care costs are rising so fast.

Our study examines the impact of the capital prospective payment system (CPPS), implemented by Medicare in 1991, on capital expenditures and cost-effective behavior of non-proprietary hospitals. As noted in the paper, we use audited financial statement data for a large national sample of hospitals. Univariate analyses demonstrate a statistically significant decline in capital expenditures in the years following the CPPS regulation without significant changes in relative aggregate operating expenses. These preliminary findings suggest that CPPS induces some cost-effective behavior by hospital managers. Ordinary least-squares (OLS) regressions indicate that capital expenditures before and after CPPS are differently affected by the changes in most explanatory variables. Further OLS regressions indicate that high-cost (low-cost) hospitals decrease (increase) capital expenditures following CPPS, once other factors are controlled for. Managerial accounting implications for hospitals include the effect of the regulation on capital budgeting decisions. Greater accounting disclosure may be necessary so that alternative modes of coping with the regulation can be discerned. Policymakers and regulators should also be aware that although reductions in capital expenditures may have favorable short-term effects of reducing healthcare costs, a potentially negative public health impact may result if capital expenditures continue to decrease.

Ran Barniv, Kreag Danvers, and Joanne Healy (2000). The impact of Medicare capital prospective payment regulation on hospital capital expenditures, Copyright © 2000 Published by Elsevier Science Inc.

CCPS is a regulation that changed reimbursement for hospital inpatient capital costs from a cost-based to a fixed fee per patient system. This apparently eliminated previous regulatory incentives to use debt and eliminated implicit guarantees of hospital debt service. I'm not quite sure what that all means except that the government implemented a program change in Medicare that altered the way hospitals do business. According to the article, the results of the policy are not clear. Legal cases involving further efforts to understand the government regulations suggest a new place to start looking for the explanation of high health care costs.

Medicare and its rules and regulations may play a role in the rising cost of health care. Medicare was fashioned after Blue Cross and Blue Shield, so these too play a role in the increasing cost of health care. All of these programs and other forms of private insurance spread the cost of expensive procedures and equipment across the population, but they do little to lower or control the cost. They are the problem and the solution. To lower the cost, we must be unwilling to buy the service at the current price. But since the expense is spread over the population and costs me little out of my pocket, I am willing to have someone else pay any price for my heart valve. The solution is not to do without the heart valve, but we must get health care cost under control.

What Have We Learned?

The result of our brief analysis is that the answer to the question "why are health care costs rising so quickly" may be found in the government rules and regulations that govern private insurance, Blue Cross/Blue Shield, and Medicare and Medicaid. The heart valve is, in fact, too expensive for you pay for yourself. The cost must be spread across the population, but at what cost? It seems that government rules and regulations in conjunction with insurance companies have allowed the medical industry to create "artificially" high prices. This is because there was an ability and a willingness to pay for the heart valve (and other medical care). Now there is a shrinking ability to pay. This explains how health care prices were allowed to rise, but it does not explain why the prices were allowed to rise.

The goals and objectives behind the rules and regulations that are passed are based in profit and politics as much, or more, than they are in an effort to control health care cost. While it may be possible to use mathematics to measure the effect of government rules and regulations on health care spending, this will not explain the reasoning behind the political decisions. The fact that politicians receive campaign contributions from some of the people that benefit from policy decisions may provide some explanation. Perhaps an important outcome of increased quantitative literacy is that politicians will be held to a higher degree of accountability for making decisions that further the cause of the people instead of furthering their effort to be reelected.

The result of the analysis is not that the application of mathematics has not provided an answer. On the contrary, based on our analysis (you should continue the investigation on your own), I believe that the mathematics has revealed the true reason for rising health care cost. The true villain is politics with the usual sprinkling of corruption and greed. The noble effort to provide affordable health care coverage to all people has been turned into a political and financial industry. The solution is for quantitatively literate citizens to demand greater accountability through greater access to data.

Summary

Our brief analysis of health care spending was not a demonstration of how to apply lots of mathematics to a situation. Instead, it is an example of how a desire to understand something together with a little thought and reasoning can produce valuable insights into highly complex and important quantitative social issues. The investigation is not complete. Whether or not you continue to investigate the issue of health care on your own will depend on your desire to know (and access to data). If you want to know more, then you will read more articles on the topic. If you find those articles lacking validity, as I did, and if you chose to be critical of poorly justified arguments, then you may desire to see and analyze the data for yourself. If these things happen, then you will have a need to know mathematics.

The alternative is to close your eyes and to allow others to make decisions on your behalf. But you must know that some of their decisions may not always be in your best interest. This situation is not unlike the person who cannot read. Such a person is at the mercy of others to interpret rules, regulations, rights, contracts, advertisements, job qualifications, and more. It is easy to understand that the person who cannot read is at a distinct disadvantage in life. It should also be clear that the person who cannot read and analyze the data for themselves faces similar disadvantages.

Some knowledge of mathematics is certainly a requirement for reading data, but it is not the most important thing. Considerably more important is your willingness and desire to be aware of and to reason about the quantitative rules and relationships that govern change in your physical and social environment. Everything changes. You may not be motivated to understand many things, but you should be motivated to understand why health care costs rise at a rate of 100 billon dollars a year. It is your money and your health. How much more personal can the issue be?

Why Do I Have to Know Mathematics?

Now, let's suppose that you have the desire to think about and reason about the quantitative aspects of change. Now, what mathematics do you need to know? Actually, this question is not unlike asking your English teacher what words you have to know. Both sets of needs will depend on what you are trying to do. Words are not only used to communicate, but they also help me think and reason. Similarly, mathematics is not just for computation; mathematics also helps me think and reason.

If you are writing a paper on health care spending, then there are certain words, some of them new, you will need to know that you might not need to know if your were not writing such a paper. If you are doing an investigation of health spending data, then you will need to know certain mathematics concepts that you might not need to know if you were not doing such an investigation. Since it is not possible to know ahead of time what you will write about or what data you will analyze, it is also not possible to determine ahead of time what words or what mathematics you need to know.

Chapter Three

Quantitative Discourse: It's All about Change

There was a time when reading was not required. Social requirements change and are changing.

In the previous chapter, we began a conversation about national health care expenditures. This was not unlike the kind of conversation you might have with friends and colleagues over coffee with the possible significant exception that the discussion centered on the actual analysis of spending data. Many people have had and will continue to have discussions about Medicare, Medicaid, insurance companies, and pharmaceutical companies. Their discussions may focus on a number of different aspects of the issue, but all such discussions are impacted by change in cost. Their discussions can be characterized by the way they choose to deal with quantitative data and change.

When discussing and analyzing data of any sort the key question is always how does the data change? Change for better or worse is at the center of most discussions. Whether it is a matter of change in views, change in price, change in status, or change in temperature, change is on our minds. Change has been the center of discussion for centuries. There have been notable discussions about the changing positions of the stars that initially placed the earth in the center of the universe and later after centuries of discussion led to our current understanding of the universe. There have been discussions of gravity and discussions about the relativity of time.

These centuries old discussions about change were started by small groups of people and grew into a whole community of scientists and mathematicians. Along the way, some business and government officials that made use of mathematics had joined the conversation, but for the most part, the group was limited to scientists investigating the nature and source of change in our physical and social environment.

First, there were small pockets of individuals working alone. Ideas were rarely shared due to distance, religion, and national boarders. As the discussion continued over the centuries and due to the development of the printing press, the scientists grew into a cohesive community with shared interests, beliefs, values, and expectations about what was important about change and what could be known about it. The community developed a special language that was and is still used to represent change and the relationships between changes. They developed a symbol system and rules for syntax. This larger group of scientists eventually spawned a smaller group of individuals that specialized their discussions to deal with the abstract representations of change. This smaller group came to be known as mathematicians. A key point here is that most of the people engaged in discussions about change were not and are not mathematicians.

When the scientists formed a special language and common beliefs, values, and expectations, they became more than just a group. They became a subculture. There are many subcultures like the scientists and mathematicians, and we all belong to several of them at various times in our lives. Subcultures form due to shared interest in music, or art, or business, or farming. There is a sports subculture, a gardening subculture, a medical subculture, a financial subculture. Each of these subcultures has its own shared beliefs, values, expectations, and special ways of communicating among its members. These special ways of communicating grow out of the activities of the group.

We belong to many such subcultures at the same time because we all engage in a number of different activities on a regular basis. When we are around others from one of our subcultures, we take on the beliefs, values, and expectations of the group. We alter our use of language by the use of specific words and phrases that are consistent with group norms. We will call these group communication practices a *discourse*.

Before we started the discussion on health care spending in the last chapter, we had asked the question: "how do I need to know mathematics?" The answer to the question is that you must know enough mathematics to be able to engage in quantitative investigations and discussions. You must develop a quantitative discourse. There are two ways to learn a discourse. You can separate the words and symbols from the activity and study the discourse in isolation, or you can engage in the activities that spawn the discourse. This is not unlike the need to surround yourself with native speakers of a language in order to truly learn the language.

To really develop a quantitative discourse, you must do what you do to learn every other discourse. You must engage in the activities of the community of the discourse. If you want to learn the sports discourse, then you have to play and or discuss sports with other sports fans. You have to become a sports fan. If you want to learn and develop a quantitative discourse, then you have to engage in the analysis and investigation of how things in your social and physical environments change. You must engage in discussions about change. The development of your quantitative discourse will come as a result of your engagement in these discussions.

In general, your need to know new words grows out of the kinds of discussions you have. When you begin to discuss new concepts and you need to express new feelings and emotions and to explain new understandings, then you will need new vocabulary. Sports fans develop special words and phrases to communicate excitement over the quality of a play. These phrases are very different from the phrases used by the theater community when discussing a good play. As communities continue to discuss sports and plays, they continue to develop their discourse. It is the same with mathematics. Your need to know mathematics will depend on the kinds of discussion you have about change. These discussions will develop your quantitative discourse and your need to know mathematics.

The idea of developing a quantitative discourse is different from the usual sense of studying mathematics found in schools today. Traditional mathematics classroom instruction focuses mostly on teaching of the mathematics language through the discussion of abstract change. Learning the language is not the same as learning the discourse. This is similar to an English classroom where you may focus on learning the grammatical rules, spelling, and definitions that are needed to communicate. But because you are constantly engaged in activities that require the use of language, you are able to use what you learn in the

English classroom every day. However, because you do not engage in quantitative discussions about change on a regular basis, you do not have the opportunity to apply what is learned in the mathematics classroom.

This is the beginning of the gap and where the seed for the questions is planted. To correct this problem, close the gap, and prevent the seed from growing, you must engage in quantitative discussions. By having quantitative discussions about change, you will develop the need for mathematics. These discussions will give meaning to the abstract ideas discussed in the mathematics classroom. The good news is that having these discussions will not require a major shift in your focus. Remember that change is all around us, and our attention is naturally drawn to it. Change is so common that it can be and is often easily taken for granted and so ignored. What there is to be understood about our physical and our social environment is centered on change. Thus developing a quantitative discourse is not far from our center of attention. We simply need to develop a sharper focus on the details of change and use that focus to develop greater understanding of our physical and social environments.

Note the real goal is understanding change, not necessarily learning mathematics. When we engage in a search for understanding about change, then quantitative relationships and mathematics become a means to an end. If there is no search, then there is no need for mathematics, but if there is no search, then we suffer the consequences of not knowing. It is this lack of understanding that determines how competitive we are in the new global economy, not how much abstract mathematics we know.

Qualitative, Quantitative, and Mathematical

It is important here to distinguish between qualitative, quantitative, and mathematical discussions of change. When we discuss change, we sometimes refer to the qualities of the change. We say that the change is good or for the best. We say that the change is unexpected or that it is predictable. We may discuss our excitement or frustration over the change. These are qualitative discussions about or reactions to change that reflect the way the change makes us feel.

These qualitative discussions can and should lead to discussions concerning how and why things have changed. When we seek to understand how and why things change, we gain the power to manage change. The quest for understanding requires that I measure change and analyze relationships between changes. This is how mathematics was developed, how it is used, and how you must learn and learn to use mathematics.

In order to understand change, I need data about the change. You cannot understand how the time and location of the sunrise along the horizon changes after one observation. You will need to make several observations and measurements so that you can measure changes. You can then use those measurements to identify patterns and begin to make predictions. You then test the accuracy of your prediction with more observations. You may repeat this observation, measurement, and prediction cycle several times or several hundred times, each time making slight modifications to your predictions and to your understanding of sunrises. As you discuss the patterns you find in your measurements, your discussion moves from a qualitative discussion of the beauty of the sunrise to a quantitative discussion about the timing and location of the sunrise.

There are no clear divisions between qualitative discussions and quantitative discussions. You may move back and forth between the two with one discussion supporting the other. It may be hard to distinguish one from the other. The primary difference is the beliefs, values, and expectations that govern your discussion as you move up and down the diagram chart shown here.

Real world Lived experience The here and now	Qualitative	Abstract Reported results
Informal hypothetical patterns and relationships	Quantitative	Rules procedures Formal patterns and relationships

Predictions		
Definitions	Mathematical	Proof of cause and
Properties		
Cause and effect		effect

The left side of the chart represents the discussions that the average citizen needs to have while the right side of the chart mirrors this pattern but from a bit more formal mathematical perspective. Discussions will move up and down either side of the chart as the discussion gets more or less quantitative. The left side of the chart is driven by real world concerns, and the right side of the chart is driven by the need for scientific precision and deductive proof. The left side of the chart is where we make decisions that impact our daily lives. The right side of the chart is where we expand our understanding of the basic nature of change and the system in which we live. The left side of the chart is where farmers develop a basic understanding of sunrise, sunset, and the planting seasons, and the right side is where scientists measure the distance from the sun to the earth, changing angles, and reflection and refraction of sunlight to explain the weather.

The qualitative discussion is based on personal needs and emotional preferences. Simple comparisons are made to determine bigger, better, proper or improper, enough or not enough, and other choices. What is good depends on the individual and his or her goals, and the beliefs, values, and expectations that drive the discussion are personal. As the discussion becomes more quantitative, there is a greater need for standardized measures and comparisons. There is more need for accuracy of comparisons. What is good is based in the application of rules and principles derived from past personal and shared community experiences. The beliefs, values, and expectations are institutional. Personal desires are not relevant. Acquisition of knowledge drives comments and gives purpose for discussions.

If and when the discussion becomes mathematical, there is a high need for precision in measurements and in thought. What is good is based on logic and the pursuit of logical arguments that provide proof of the correctness of predictions. The beliefs, values, and expectations that guide the discussion are based solely in the domain of the human ability to know and develop an objective understanding of the universe.

A mathematics discussion is always just beneath the surface of any quantitative discussion. You can have a qualitative discussion about simple and compound interest that leads to a quantitative discussion and comparison of investment options, which can lead to a desire for greater understanding of the concept of interest. As this discussion progresses, it may move from qualitative to quantitative. Most people do not need to develop a highly practiced mathematical discourse. They do not need to spend much time on the right side of the chart. However, the more your discussions about change moves up and down the left side of the chart, the more likely it is that you will begin to move left and right as well as up and down the chart.

A Discussion about Change?

Always keep in mind that everything changes. Therefore, this discussion applies to everything. However, you should also remember that the key to any investigation about change depends on your interest in the change. If you do not want to measure, predict, and manage change, then you are not ready to engage in quantitative discussion, so you will not develop a quantitative discourse. But if you are not ready, then you are limiting your ability to compete in today's global knowledge economy. You limit your ability to raise questions about health care spending and every other aspect of politics and the economy. You limit your ability to control the changes in your life.

Every discussion of change begins in the same place, with a simple comparison of two or more numbers. There are simple and ordinary aspects of change (the comparison of the two numbers) that become important when having quantitative discussions. First, we note that change can be fast or change can be slow. Things can change so fast that you can't see the change, or they can change so slow that you can't see the change. The shutter on a camera changes position in the blink of an eye while the hour hand on the clock moves slowly. Both are difficult to see. In Phoenix, Arizona, ice in a cup of water melts more slowly than ice in the middle of the street. Fast and slow is a measure of the amount of time required for an object to move from one position or state to another.

A significant amount of mathematics deals with measuring how fast or slowly things change. Understanding how fast or slowly something changes also occupies a significant part of our thoughts about issues in our physical and social environments. How fast is the economy changing? How fast is the temperature changing? How fast is the rookie quarterback's understanding of the game changing? So we are already engaged with the central topic of mathematics. We are already interested in change. So in a sense, we do not have to do anything new to develop our quantitative discourses; we simply have to do more of what we already do.

The reason that our interest in change has not already led to a natural understanding of the importance of mathematics is that our interaction with change is typically of a superficial qualitative nature. We evaluate change as good or bad based on our ability to deal with the change. We evaluate change as tolerable, necessary, or sufficient. We typically deal with change on a case-by-case basis, as a concrete situation, and we consider one change at a time. We ask how much gas prices have gone up today, but we do not keep accurate records of price changes from day to day. We don't ask what has been the overall trend in gas prices in relation to the growing economy in China. We may see the change in gas prices as an indicator of a systematic process that is being be managed and controlled; however, if we do not seek to understand the system, then we will not be able to manage it as we should.

Our superficial interactions with change do not support the development of quantitative discussions. Quantitative discussions are situated in an effort to develop understandings about relationships between two or more changes. Quantitative discussions situate a simple change in a larger system of changes and in an effort to understand why the change is fast or slow. For every action, there is an equal and opposite reaction. Thus, a change in gas prices was caused by a change and will cause a change. Quantitative discussions are about understanding the system of changes.

Change can also be big or small. When we measure the size of a change, we have to compare the new amount to the original amount. If I tell you that there are six more apples, then while you have some information, you are missing an important piece of information that is necessary to be able to place the six apples in a proper context. What is missing is the original number of apples. If there were originally three apples, then six more takes on a different meaning than if there were originally 350 apples. In either case, there are still six more apples. This is the basic idea behind percent change that we discussed in chapter 2. Going from three apples to nine apples is a bigger change than going from 350 to 356 apples, even though there are still just six more apples.

Change can be patterned or random. The weather follows predictable patterns. Most things in our lives also follow predictable patterns. We like changes that follow predictable patterns, so we try to create similar patterns. We eat at the same time everyday, and we wake and sleep at the same time. We develop a daily routine. As things become less and less predictable, they become more and more random.

We are all accustomed to the changes in the seasons. When change follows a predictable pattern like summer, fall, winter, spring, we expect the pattern to continue. Of course, when the seasons change depends on where you live, but when the latter days of September begin to cool, it is expected. If this change happens early in September, we may be caught a bit off guard, and we may be a little concerned. If the days do not cool until late in October, then we are pleased; however, we still note the change in the pattern of the usual beginning of fall weather. Some argue that due to global warming our weather patterns are changing, and others argue that the changes in our weather are just normal random fluctuations. The point is that we all notice the change, and we all discuss the weather. To develop the quantitative discourse you need, you have to expand your discussion of these changes.

Change can be constant or nonconstant. When we drive our cars, we experience both constant and nonconstant change. When we set the cruise control at 55 mph, then we are experiencing a constant change. Our distance from our starting point and to our ending point is changing at a constant rate of 55 mph, and our speed is how fast our position is changing. If we in fact have 82.5 miles to go, it will take one and a half hours to reach our destination. When we change speeds by either hitting the breaks or the accelerator, we experience a nonconstant rate of change. When you are accelerating or breaking, it is difficult to say how fast you are going since your speed is changing at each instant—that is, your speed is nonconstant. This is like trying to say where a fly is that has not landed. It's hard to be specific.

It is obvious that a change can be either an increase or a decrease in value. Thus we can have something that always increases. Time always increases relative to some starting point. We can also have things that constantly decrease. These are some of the ways things can change, and they are not new to you. You already use these ideas on a regular basis to think about the changes that you experience every day. Here are some of the changes that you deal with every day.

Wages	The economy	The poverty line
Tax rates	Cost of gas	Cost of food
Cost of a home	Your health	Price of milk
Foreign relations	Foreign markets	Attitudes
Politics	Opportunities	Price of stocks
The value of a dollar	Your weight	Health care cost

This short list of things that change is meant to help to emphasize that the need for mathematics really is all around you. You have already taken the first step toward realizing the power of mathematics when you recognize and think about changes in your physical and social environment. To complete the process you have to begin to think more systematically about the changes that you experience. Every change was caused by a change and causes another change. To complete the process you must seek to understand the system of changes.

Interpreting Change

The first thing we must do to begin a discussion of change is to measure the change. Measuring change always involves the comparison of two numbers. To determine change, we usually subtract the first or initial value from the second value. The difference between the two numbers is the actual value of absolute change. It is the size of the change.

A simple difference is sufficient in many situations and is as we said always the first form of measurement used in the study of change. The need for more complex measures of change depends on what you are trying to understand. Thus, as always, your need for mathematics is driven by the kinds of questions that you ask. This book is as much about getting you to ask the questions as it is about helping you to understand why mathematics is so important. The point is that I can tell you why mathematics is important, but if you do not ask the questions, then you will not understand the answer any more than a standard answer like, "mathematics is all around us."

Once you have the simple difference in values you have to decide what the simple difference means. How do you interpret the simple difference? For example, if you are 21 years old, and you currently make sixteen dollars an hour and work 40 hours a week, then your base pay is $640.00 a week. If you get a fifty cent per hour raise, then your base salary will increase to $660.00 a week. In terms of absolute difference, you make $.50 more an hour, $20.00 more per week, and $1040.00 more per year.

The first step is complete. We have measured the difference. How we interpret the difference depends on a lot of things and most of them have little to do with mathematics. That is, mathematics has little to do with determining how you interpret your raise. The hammer, wood, and nails have little to do with the design of a house, but most houses cannot be built without a hammer, wood, and nails. So while mathematics does not determine the questions you ask, it will certainly play a central role in answering the questions. How much mathematics you need and the role it plays depends on how big the house is that you want to build (or how much understanding you want to develop).

As you begin to interpret your raise, you may start by considering your current financial situation. What is the actual impact of the additional twenty dollars a week on your ability to purchase necessities? Was the original sixteen dollars an hour enough to pay your bills, or were things already a bit tight? There are practical, in your face, observations of your existing situation that on the surface do not require much mathematics to understand. Your first qualitative interpretation might be that the raise is small and therefore not enough to make a significant difference in your economic situation. You may not even see the need to employ the powers of mathematics to determine the impact of twenty dollars a week on you financial well-being. But this is where your desire to understand comes into the picture.

Why would you want to know more? Why should you invest any more time in trying to understand the economic relevance of your twenty dollars a week raise? Simple curiosity is one reason, but the ability to make more informed decisions that will benefit you and your family is a better reason. A greater understanding of your raise and how it fits into your local, state, and national economy will help you make better decisions and will make you better able to judge the decisions being made by local, state, and national elected officials who are supposed to be acting in your best interest.

It is your desire and need to understand the larger economic system that moves you beyond the basic absolute value of your raise and or your ability to afford the expanded sports package from the local cable company. Let's enlarge the context a bit and see how your raise is related to other important factors. You and your hourly wages are a part of a much larger economic system that is controlled by business and political rules and regulations, as well as by human need and greed. Your twenty dollar a week raise may allow you to maintain your current level of existence, or it may not be a raise at all. In fact it may ultimately reflect a loss of buying power. It is only by asking questions about and seeking to understand the nature of your raise that you gain the power to manage your economic situation in your best interest.

How does your raise compare to the increase in the cost of a loaf of bread, a gallon of milk, or a gallon of gas? The consumer price index (CPI) provides a way to understand your raise in terms of your local, state, and national economy. Perhaps the easiest way to describe the CPI is that it is a measure of the how much you can buy with a dollar. The CPI is computed and reported by the Bureau of Labor

Statistics (www.bls.gov) each month. According to the BLS, "The Consumer Price Index (CPI) is a measure of the average change over time in the prices paid by urban consumers for a market basket of consumer goods and services." If you get a 3% raise and the CPI goes up 2%, then you really only got a 1% raise because it will cost 2% more to purchase the things you need. None of this is exact. The CPI may be an accurate measure of cost changes in your area, or it may not reflect prices in your city very well at all. However, it does provide a fixed reference point for making comparisons. Comparing your raise to the increase in the CPI will help you to understand how you are doing in the economy.

Again this simple comparison can be the answer to your final question, or it can be the start of a larger investigation. How is your city and state doing with respect to the rest of the economy? How is your company doing with respect to the economy? What are your company's profits? Are your company's profits in line with the profits of similar companies? How much does your company have to contribute to cover health care benefits? How much of that money would go to salaries instead of health care if we demanded to know why health care spending has gone up so fast and so high?

The CPI

Knowing that the CPI for December 2008 was 210.228 tells us how the value of a dollar has changed. To measure change, we already know that you have to have a starting and an ending value. In the case of the CPI, the starting value is currently the average value of a dollar in the years 1982 to 1984. This average is set as 100, and all changes are measured relative to this value. The December 2008 index means that it cost $210.29 to buy what $100 bought in 1984. This is an absolute difference of $110.29 and a 110% change.

To understand your raise in terms of the CPI, first we must pick a starting year. Knowing how much buying power you would have had in 1984 is informative, but it is not the information you currently need. What you need to know is how the value of the money in your pocket is changing now. The CPI for December 2009 was 215.949. If we subtract the December 2008 CPI from the December 2009 CPI and divide by the December 2008 CPI, we will get the percent change in the CPI. If we compare that percentage to the raise you got in December of 2009, we can tell if you are doing better, the same, or worse relative to the economy.

$$\frac{(215.949-210.228)}{210.228} = \frac{5.721}{210.228} = 0.0272 = 2.72\%$$

If we compare the 2.72% increase in the CPI to the 3.0% increase in your wages, we see that you got the equivalent of a 0.28% raise. Thus after accounting for inflation, you have a net four cent per hour raise. You are actually fortunate to be keeping pace with inflation, and no, you cannot afford the expanded sports package from your cable company.

Consider the following example. If your salary was $35,000 in 1995. Then suppose that you got a $5000 dollar raise in 2000 and another $5000 dollar raise in 2005. On the surface, it would appear that you are making $10,000 a year more than you were in 1995. However, things are more expensive in 2005 than they were in 1995. A dollar in 1995 bought you more than a dollar in 2005, so it takes more money in 2005 to do the things you did in 1995. Do you really make more money? To find out we must convert 2000 dollars and 2005 dollars into 1995 dollars by using the following equations.

$$\frac{1995: CPI}{2000: CPI} x2000: Dollars = 1995: Dollars \text{ or } \frac{152.4}{172.2} x\$40{,}000 = \$35{,}400$$

and

$$\frac{1995: CPI}{2005: CPI} x2005: Dollars = 1995: Dollars \text{ or } \frac{152.4}{195.3} x\$45{,}000 = \$35{,}115$$

You can see from the calculation that your raise in 2000 was the equivalent of a $400 dollar raise. That is, in the year 2000, by the time you purchase everything that you purchased in 1995 for $35,000, you would have $400 left over. Also, you can see that in 2005 your $5000 raise actually resulted in a decrease in buying power. At this rate, you will never be able to afford the expanded sports package from the cable company or a college education for you or your children.

Consider a mortgage payment in 1995 that was $1000 per month. In 2000 that same $1000 payment would be equal to $1128.99, and in 2005 it is equivalent to $1281.49. However, you still only have to pay the same $1000 per month. The result of having a fixed mortgage payment is that in 2005 you have earned a $281.49 per month or $3377 a year raise. That sports package is looking like a possibility. So there are ways to do battle with the CPI monster that we have all created.

Consider the graph of the CPI from 1913 to 2009 in figure 3.

Figure 3.

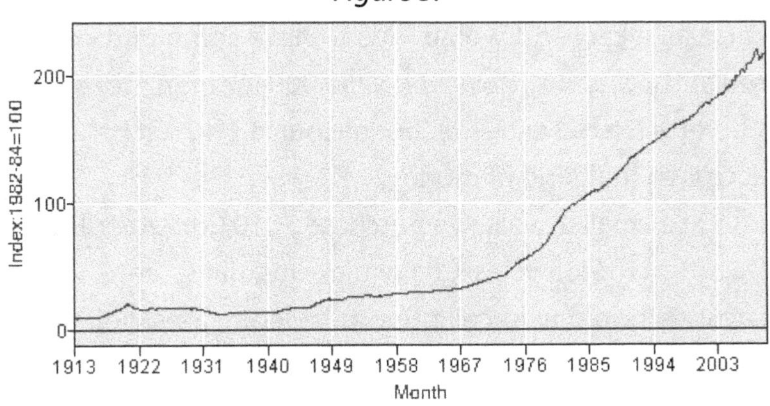

What questions do you see? What you should notice in the graphs is the sharp rise that takes place around 1976, and then your next question might be, why? The result of this sharp increase is that a $5000 raise is really a

$400 dollar raise and your 50 cent per hour raise is really a 4 cent per hour raise. You may be too young to remember the double-digit inflation period between 1970 and 1980. Interest rates rose to 10% and higher. Inflation is not a good thing. How do we control inflation? We can wait for the government to control it for us, or we can help by doing our part. To do our part we must increase our understanding by asking questions and by analyzing data. We need mathematics.

What causes inflation? The answer is simple—you do. When people have easy access to money, they are willing to pay higher prices for the things that they want. The Barrett Jackson Auto Auction is an example. There, a room full of people with way too much money bid on vintage cars. I like cars and would love to have some of the cars I see at the auction; however, $250,000 (or more) for a 1966 Corvette is a bid extreme (but of course, I do not have easy access to that kind of money).

On a smaller scale, we purchase 2010 Corvettes, Harley Davidson motorcycles, Chevy pick-ups, flat screen TVs, the latest technology in cell phones, video games, golf clubs, movie tickets, and a number of other items because we want them. The price that we are willing to pay for these items is a source of inflation. There are also necessities like gasoline, heating oil, food, and insurance that we must purchase that play a role in inflation. All of these prices are managed by a number of people, forces, and institutions so as not to bring about a collapse of the economy. Virtually no one wants the economy to collapse. Some manage the system for their person gain, and others attempt to manage the system to protect you the consumer. However, the primary manager of the economic system and inflation should be you, the individual consumer.

Producers conduct market analysis to determine the best selling price for their goods. Their goal is to maximize profits, which opens the door to selling their product for much more than it is "worth." This is because the value of the product depends on the subjective price that you are willing to pay and not on some objective value of product. It is not even clear that there is a way to assign an objective value to something like a cup of coffee or a laptop computer. When producers set their prices, they consider their production and distribution cost and then try to sell their product for as much as your will pay. Did you know that if a high-quality product is priced too low, people will not buy the product for fear that it is inferior? So producers are actually encouraged to set higher prices.

There is a limit as to how high they can set the price. You will purchase a standard issue ballpoint pen for $3.89, but you will not pay $5.00 for the same pen, especially given that you have other options. If gas prices rise too high and people can no longer afford to fill their tanks, then they stop buying gas and profits fall. If prices are too low, then companies don't cover their costs and are eventually driven out of business. So there are a number of factors that must be balanced to insure that the system is fair to producers and fair to consumers. When one player understands the rules of the game more than the other, then there is a clear advantage for the more knowledgeable competitor. As a rule of thumb in business, you do not want to concede such an advantage to your competition. Consumers have made this concession, and we pay a heavy price for it every day.

The choices that you make concerning what you do with you twenty dollar a week raise, and the rest of your salary too, have a significant impact on the economy. This is why it is critical that you understand how your local, state, national, and now international economy works. We often feel that we have no control over the economy because it is too big; however, you must understand that every dollar that you spend impacts the economy. Every time you spend that dollar without understanding the impact of your purchase, you give up some of your rights and shirk off some of your responsibilities as a citizen.

The movie industry represents a clear example of the power of your purchasing decisions. You may use you twenty dollar raise to purchase a movie ticket for you and a date. Your night at the movies could cost you $40 to $50 dollars if you throw in popcorn and a soda for two. Why? Do Mel Gibson, Will Smith, Demi Moore, or any other actor or actress need or deserve twenty million dollars for a movie? What do you have to say about how much money they get paid? Perhaps they are only able to command these payments because you are willing to spend $10 or more per ticket to see the movie in the theater. When an actor or actress makes a movie that does not do well in the box office, he or she may not be able to demand as much money for the next film. And the reason that the film did not do well is because you chose not to go and see it. Thus, you see the power of your choices as to how and where you spend your money.

The entertainers, manufacturers, and retailers that are successful owe their success to your choice to purchase their services. Everyone seems to understand this fact except the most important part of the economy. The ones with the least informed view of the whole process are the consumers, and yet we are the ones with the most to gain and the ones that make the whole thing work.

We have recently experienced an economic collapse that has many recalling the days of the Great Depression. The central cause of this collapse was the collapse of the mortgage industry, which was caused by greed and a lack of mathematical understanding. It is important that we learn from this mistake as we move forward. How do we avoid similar shortsightedness?

The economy is driven by people who buy things, and people who buy things need money. Low interest rates make it easy for people to get money. When it is easy to get money, people are able and willing to pay more for the things they buy. When it is hard to get money, people don't buy and prices are forced down.

Of course, these same people who make the purchases that keep the economy running also work for companies and corporations that produce the goods and services that we purchase. Thus what we seek is a balance, where consumers are able to purchase the things that they need and the things that they want, and where producers are able to make a "fair" profit.

The analysis of your twenty dollar raise could lead to a better understanding of your local economy, which would lead to smarter spending decisions, which would ultimately maximize the impact of your raise. If you feel powerless to change things, it may be because you do not understand how things work. If you do not understand how things work, then you do not know the power that you have to effect change. Mathematics gives you the power to know both how things work and to recognize and know your ability to make change. To use the power of mathematics, you must ask questions concerning how things change.

The point here is that there are a number of quantitative discussions that could result from an investigation of the relevance of your raise. You should engage in these investigations whether you get a raise or not. It is only by understanding the quantitative nature of your social and physical environment that you will be able to exert control. You may ask: where is the mathematics? The answer is that the mathematics is in the questions. I am asking some of the questions that need to be asked. If you do not ask these or similar questions, then you do not need to know mathematics. If you do not ask these questions, then you will continue to place your fate in the hands of businesses whose only goal is to make more money for their investors, or in the hands of politicians who owe too much of their political success to these same businesses. These politicians should owe their success to you.

You don't learn to read simply so that you can read novels and comic books. By learning to read, you also posses the power to read newspapers and Web sites that provide news and information on politics, business, the economy, and other important areas. By reading, we stay informed. When we are informed, we can make better decisions. It is your right and your responsibility as a citizen to be informed.

Business leaders and political leaders do not make decisions without first being informed, and this means seeing the numbers. The numbers help them to understand the long- and short-range impact of different choices. If you don't also consider the numbers, then you are not truly well informed? Why do we only want to be half informed? Why don't we also want to see the data? Why don't we want to read the numbers? This is why you have to know mathematics, so that you can be fully informed.

As you have seen the government collects lots of data that and gives the public free access to use the data. Every government agency collects data, and every business and company collects data. Sports teams and movie producers collect data. TV and radio stations collect data. They all collect this data so that they can be better informed and make better decisions. Those who have access to the data have access to power. Those who can and do use mathematics to understand what the data means have power. If you do not have access to the data, then you do not have access to the power. If you have access to the data and choose not to understand the data, then you give your power to others who will use it for or against you as they see fit.

Investing

Perhaps, there is no place in your social environment where access to data is more important than in the world of business and investment. Assuming that your bills were covered by your original salary, you could decide to invest your twenty dollar a week raise. Investing is another way to hedge against inflation. Deciding to invest the raise will lead to a need to know and understand the concept of compound interest, the formulas used to compute interest, and the concepts of linear and nonlinear growth. You can trust someone else to manage your twenty dollar a week investment for you, or you could chose to understand your options for where and how your money is being invested.

Many people assume that twenty dollars a week is not enough money to invest. There may be many reasons for this. I offer two rather obvious possibilities. First, the news seldom reports investment stories centered around an individual who has invested small amounts of money. Instead, the news is about the millionaires and billionaires who make, lose, or steal large sums of money. Second, most people are not thinking long term when it comes to their economic situation. Many people are concerned with making ends meet on a day-to-day basis and so do not take the time to think about life 25, 30, or 35 years down the road. However, if they would take the time to think that far down the road, they would see that their twenty dollar a week raise could potentially make a significant difference in their futures.

Your twenty dollar a week raise equals a $1040 a year raise. If each year, for the next 30 years, you were to place your current $1040 a year into an investment that pays 5% annual interest, the results would be significant. However, you will probably continue to get raises as the years pass. You could also choose to invest all or part of those raises, which would increase your annual contribution to your investment fund. We will see a little more about this later.

For now however, let us assume that you stick with the $20.00 a week or $1040 a year investment. In thirty years, you will have placed $31,200 dollars of your money into the bank; however, the investment will be worth $69.096. Over the course of thirty years, your investment will more that double in value. If you continue to make the same investment for five more years, then the value of the investment increase by another \$24,836 to be worth a total of \$93,933 or almost \$100,000.

The question you have to ask when you are 21 years old is do you plan to be alive in 35 years and would you like to have \$93,933 on your 56th birthday. If you decide to wait until your 61st birthday to spend your investment, your money will grow by another \$31,698 to be worth \$125,693. Let's just assume that you work to 67 years old when you retire and begin to collect Social Security (if it is still around), and then your money will grow another \$49,800 to be worth \$175,432. The decisions you make when you are 21 years old about a \$20 dollar a week raise could result in a sizable retirement gift to yourself when you are 67 years old.

As I stated earlier, you will probably get additional raises either on the same job or by changing jobs over the next 30 to 35 years. You could choose to invest all or some of these additional raises into your investment fund. If you decided that every year you would invest an additional \$200 dollars into your account until in year 16, you were investing \$4000 a year into your retirement found. Then you continued to invest the \$4000 a year until you are 67 years old. Then in the same 46-year period, you will have invested \$160,600, and you will have accumulated

$504,299. Your twenty-dollar raise could be the start of something big, or you could order the NFL package on cable TV.

The idea of investing for retirement is certainly not a new idea. You may have even engaged in many qualitative discussions about if and when you will be able to retire and how much money you will have. It is time to have a quantitative discussion about your economic future. You can go to a good retirement adviser and get the same story with more information and detail than I have provided here. But you must begin the quantitative discussion of about your future. The key issues in that quantitative discussion will be time and compound interest. Time and compound interest are the reasons your investments grow the way that they do. Notice that in Table Three, if you stick with your original $20 a week, $1040 a year investment, you're your money grows to $69,096 in thirty years, to $93,933 in 35 years, and to 125,631in 40 years, and to $217,721 in 50 years.

Table Three

Years	Total investment	Value of investment	Total interest	Additional growth
35	$36,400.00	$93,933.12	$57,533.12	$24,836.72
45	$46,800.00	$166,088.16	$119,288.16	$40,456.40
50	$52,000.00	$217,721.92	$165,721.92	$42,289.35

What the numbers in Table Three are showing you is that the longer you continue to make your investment, the faster your money grows. Consider the last column in Table Three. Your investment grew by $24,836 in the five years between 30 and 35, but by $31,698 in the five years between 35 and 40, and then by $40,456 in the next five years. Your money is growing/changing faster. This is the effect of compound interest, and it is an example of non-linear growth.

This example shows that the longer you leave your money in the investment the faster it begins to grow. This is where time and being 21 years old when you start to invest come into play. If you start when you are 21, then you have 46 years before you are 67. If you start when you are 31 years old, then you only have 36 years to work with. The longer it takes you to start, the less time you have for the money to grow. The result is that it is even more important that you choose to invest your $20.00 a week raise instead of spending it on a new broadband cell phone contract or the NFL sports package from your cable company.

You cannot increase the amount of time that you have; thus, the only way to increase the amount of money that you have at age 67 is to increase the amount that you save or to increase the interest rate. A 1% increase in the interest rate from 5% to 6% will increase the value of the investment by $60,135 to a total of $235,568 in the 46th year. At 7% interest compounded annually, the same investment yields $83,453 more for a total investment value of $319,021. In addition to increasing the interest rate, you can increase the number of compounding periods. This will lead to even larger gains. Are you asking questions yet?

Developing an investment strategy for your retirement requires that you understand how the value of your investment changes. If you give your money to someone to invest for you, then you are not able to judge the quality of service that you are receiving, and you are not able to manage your investment.

When you are twenty-one, you should decide how much money you will want to have in your investment by age 50, 55, and or 65. Then you should figure out how much you will need to invest each year to reach your goal. One thing that you should understand is that the more you put in early, the faster it will grow later. That is, you could start by contributing more to the investment early in life and possibly slowdown later in life. If you reverse the typical pattern of investing less when you are younger and more as you get older to invest more when you are younger and less as you get older, you can increase the net value of your return while investing less money in the long term. This means that you will have access to more of your money for spending as your get older. In actuality, when you are young is when you actually need less of your income.

Are you asking questions yet?

Chapter Four

Cause and Effect

The goal is not to teach you mathematics but to help you understand why you need to learn mathematics. With this knowledge, you will find learning mathematics to be a much more rewarding and fulfilling experience.

It is the desire to understand and manage change that sets the wheels of mathematics in motion. Every effort to understand change begins with a simple measurement or a series of measurements. Initially, these measurements may be a simple recognition of a change in speed, size, or quantity. As we attempt to developed greater and more precise understandings of change, measurements will be made using more precise instruments and scales. But these measurements alone are only numbers that must be placed in a greater context to be interpreted.

> Knowing that 5000 barrels of oil a day are spilling into the gulf is not much information by itself.

Why Do I Have to Know Mathematics?

The purposes and goals that shape our quests for understanding of how things change will help to shape the contexts that we develop for understanding measurements. It is the combination of purpose, goals, contexts, and measurements that motivates and shapes our need to know mathematics.

At the core of the contexts in which we interpret change is the cause and effect relationship. For every action, there is an equal and opposite reaction. Every change is caused by a change, and every change causes a change. To understand a change, we must know what caused the change and the effect of the change.

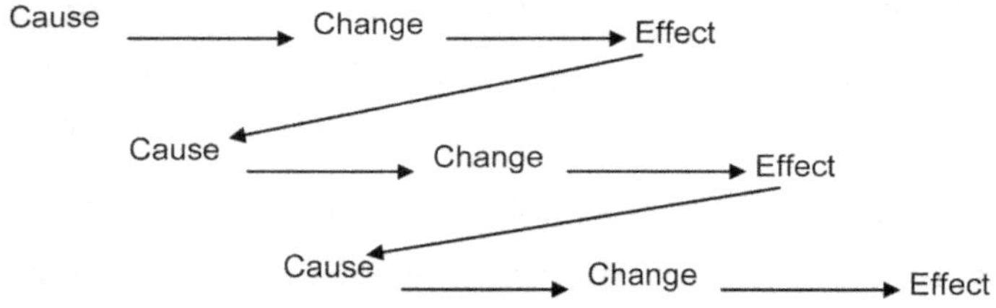

The effect of a change will in turn become the cause of another change. A series or chain of cause and effect relationships makes a **system** of changes. We use **systems** to organize our understandings of our various environments. We have a solar system, a weather system, an environmental system, a circulatory system, a health care system, a financial system, a political system, and many, many more systems. Each system consists of a number of related parts and the change relationships that exist between the parts.

Our understanding of a particular system depends on how well we understand the effect of a change in one variable on the other variables in the system.

Systems may start off small, with one or two cause and effect relationships. As we begin to consider a wider range of cause and effect relationships, our systems grow and so our understanding grows. We may start with a simple relationship, such as it rains when clouds change from white to grey. However, as we continue to observe and experience weather, we will notice that our initial cause and effect relationship does not always hold true. Sometimes the clouds turn grey and it does not rain, and sometimes it rains when the clouds are not grey. As we attempt to improve our understanding of the cause of rain, we may consider other changes in things such as temperature, barometric pressure, and humidity in addition to the color of the clouds and why the color appears to change.

As we consider more variables and more changes, our system grows and may become more or less complex. A change in humidity may be caused by a simple rise in temperature, or it could be caused by a more complex interaction of several related factors. As scientists have studied weather over the years, they have developed a complex system of variables like wind speeds, and wind currents, high pressures zones and low pressure zones, barometric pressures, dew point, cloud cover, and many more variables. Scientists measure changes in each variable in different places and use this data to determine cause and effect relationships that allow us to predict the weather. As our understanding of the system grows, our ability to make more accurate predictions grows as well.

When we analyzed health care spending, we were searching for a cause and effect relationship that would explain why health care costs are rising so fast. When gas prices rise, we want to know the cause. Whenever our attention is drawn to a particular change, it is natural for us to want to know what caused the change. Our natural drive to measure change together with our natural desire to understand cause and effect relationships are the foundation for the development and use of mathematics knowledge.

While attention to change and the desire to understand cause and effect relationships may be natural, they must be nurtured to be sustained. Imagine the world without the printing press. The ability to read may not be nearly as important as it is today. The ability to easily produce and distribute written texts makes the written word a major means of communication in our society, and this makes the ability to read an essential skill. The point is that it is the availability of reading material motivates, shapes, and drives our need and desire to read.

In order to sustain our natural desire to understand cause and effect relationships, we need to develop a national quantitative discourse community that parallels the literacy community. The literacy community consists of authors and readers. The quantitative discourse community also needs to consist of authors and readers. To support the development of the quantitative discourse community, we need greater access to data and data analysis. Investigating cause and effect relationships requires a series of measurements collected over time and in varying conditions. The average citizen does not have the resources and/or the time to collect this data. This is the job of the quantitative authors. The good news is that in many cases the government and/or private industry are already collecting the data. There are also already some quantitative authors, but they are not writing for me, you, or the general public.

There is a major reason why data and data analysis are not made more readily available to the general public. The general public currently has not expressed any significant desire to know. Instead the public has settled for qualitative statements that stir emotional and sometimes irrational conclusions. In order to change this situation, we must expect quantitative analysis in place of qualitative sound bites. We must seek to understand the behavior of our complex social systems. This is where the need to know mathematics and to develop a quantitative discourse is based. Your desire and need to understand your physical and social environments are all of the motivation you need. Your understanding of why mathematics is so important is based on your efforts to develop and understand the systems of change that are all around you.

As I have pointed out several times already, there is a relationship between the desire to know and the willingness to think, reason, and understand. An increased willingness to think, reason, and understand may lead to an increase in desire to know, but an increase in your desire to know will always support the development of a greater willingness to think and reason. Thus, your entry point to the study of mathematics and to the development of a quantitative discourse is not just through the mathematics classroom. You must also learn mathematics and develop a quantitative discourse through your efforts to acquire understandings of the cause and effect relationships that underlie the systems that make up your physical and social environments. Your efforts to understand provide fuel for your drive and desire to learn in the mathematics classroom.

Mathematics Models

A mathematical model is a precise quantitative description of the change relationships in a system. When we think of a model, we typically think of a smaller physical copy of an object like a model train or a model car. A mathematical model does not model the physical object; instead, it models the changes in the object. A mathematical model is more like a story than a physical model. In fact, a mathematical model can be written in plain everyday language like a story. As a simple example consider the equation y=5x+2. The equation expresses a relationship between the change in x and the change in y. In plain language, this statement says that the value of y can be determined by multiplying the value of x by five and then adding 2. It means that a change in x causes a five times greater change in y. Consider the following sign.

Boat Rentals

$5.00 an Hour

$2.00 Launch Fee

How much will it cost you to rent the boat for 3.5 hours? In this case, a change in rental time causes a change in rental price. The equation y=5x+2 does not model the boat or the boat ride, it models the effect of a change in rental time on rental cost. The equation tells the story, and it has meaning because of the boat rental business. Your need to know the equation depends on your desire to understand the various possible costs for renting a boat.

We use the word "dog" to represent our shared experiences with a certain kind of animal, but the word is not a dog. A mathematical equation is not a cause and effect relationship, but we use the mathematical equations to represent our shared experiences with particular kinds of cause and effect relationships. If you have never seen a dog or if you have had limited experiences with dogs, then the word "dog" will have equally limited meaning for you. Similarly, if you have had little experience measuring change, analyzing data, and identifying cause and effect relationships, then mathematical equations will have equally limited meaning. It is understanding the cause and effect relationships in our systems that gives meaning and purpose to mathematical formulas, equations, and procedures. The mathematics takes meaning from your efforts to understand and explain cause and effect relationships.

You can learn a list of animals before you go to the zoo, but if you never go to the zoo, the list is dead and lifeless. Once you get to the zoo, the list comes alive. If you go to the zoo before learning the list of animals, then the words take meaning directly from your experience. You can learn a list of mathematics topics and then try to understand things, or you can try to understanding things and develop meaning for the mathematics from the experiences of developing understanding. You clearly need to do both. You have to learn the vocabulary and go to the zoo. You have to learn the mathematics and develop greater understanding of the systems in your environment. However, a visit to the zoo motivates the need to know the vocabulary in the same way that an effort to understand systems will motivate the need to know mathematics.

Different Kinds of Cause and Effect Relationships

What causes gas prices to rise? What is the relationship between gas prices and heating oil prices? What causes food and clothes prices to rise? What causes a headache? What causes cancer? What causes one team to win and another team to lose? What causes illegal immigration? Can you list ten questions like this? Was it easy or difficult to come up with ten?

Qualitative Descriptions of Cause and Effect. We deal with cause and effect relationships all of the time. However, we typically deal with these relationships qualitatively. Observations and experience tell us that more of one thing causes more of another or that more of one thing may cause less of another. For example, the more water there is in the pot the longer it takes to boil. We also know that as the price of gas goes up then the demand for gas goes down. We do not do experiments to determine how much longer it takes two cups of water to boil than one cup. We don't measure the demand for gas when there is a price change. Instead, we allow more time for cooking if we have to boil more water, and we make fewer extra trips in the car as gas prices go up.

When more water takes longer to boil, we say that there is a *direct causal relationship* or just a *direct relationship* between the amount of water in the pot and the amount of time it takes to boil the water. When higher gas prices cause a decrease in demand, we say that there is an *inverse causal relationship* or just an *inverse relationship* between gas prices and demand for gas. If you think about it, it makes sense that adding more of component A can only have one of three effects on component B. More of component A can cause an increase in component B (a direct relationship), a decrease in component B (an inverse relationship), or no change to component B at all (no relationship).

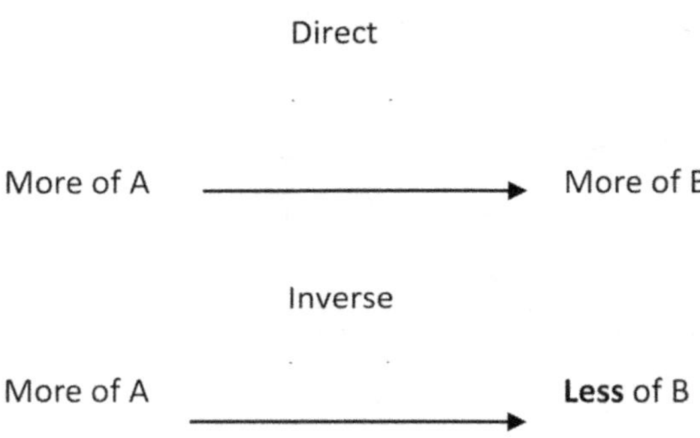

This seems like a common sense idea, and it plays just such a role in our everyday interactions with our environment. That is, when we add more of component A, we expect one of three things to happen.

We spend a great deal of time trying to manage direct and inverse relationships. The more food I eat, the more weight I gain. The more things I have to do, the less time there is to get it all done. The harder I try the worse things get. The more mass a planet has, the stronger the gravitational pull it has (well, some people think about that last one). Direct and inverse relationships are all around us. We expect them, we accept them, and we live by them. But we do not typically measure them accurately. We simply make small corrective adjustments until we get the results we want.

This is the natural order of things. Nature does not require animals or humans to understand. Our political, economic, health care, and other systems also do not require you to understand. The order that is provided by natural and man-made systems allows us to manage our environments by making small adjustments to changes that occur, thus, giving us the illusion of control. But we cannot control what we do not understand. If we do not understand the weather, then we cannot predict, manage, and perhaps ultimately control the weather. We cannot prepare for droughts or hurricanes. To improve our ability to predict and manage the weather, we need more knowledge of the factors that bring about changes in the weather. We need to measure the effect of humidity on temperature and the cause and effect relationships between weather, rain forest, and ocean currents, farming, cars, and factories.

If we want to understand any system, we need to understand the cause and effect relationships that exist between various parts of the system. Qualitative measures will only allow a limited level of understanding. To go further to the point where we can predict and manage the system, we must collect and analyze measurements of the cause and effect relationships in the system. We must use quantitative methods.

Quantitative Descriptions of Cause and Effect. If we want to have a better understanding of the relationship between the change in component A and the change in component B, then we need data. We can make detailed measurements and observations of natural events as they occur, or we may be able to conduct a series of experiments by adding different amounts of component A and then measuring the increase or decrease in component B. This is a natural process of everyday observations and trial and error that we improve by making accurate measurements and recoding data. Once we have a series of data measurements, we are able to look for patterns in the changes and describe the relationships that exist between changes in the components of the system. We can use our measurements to make predictions and determine exactly how much of component A to add in order to get a specific amount of component B.

If the change in component B is always the same for a given change in component A, then there is a strong relationship between the two components. We can make accurate predictions. Nature provides lots of these relationships. The way that heat, gravity, and electricity interact in systems produces predictable cause effect relationships.

In the case of heating water, if we collected data from experiments that measured the amount of time it takes to heat various amounts of water, we would find that under consistent conditions there is a very exact relationship between the volume of water and the time it takes it to boil. By conducting many such experiments scientists have determined that changing the temperature of one 1 kg of water by 1°C requires 4186 J (or 4186 watt seconds). If our heat source provides 1200 watts (like an electric stove), then it takes 3.488 seconds to raise 1 kg of water 1°C. If the water started at 25°C (room temperature), we would need to raise the temperature of the water by 75°C to 100°C the boiling point of water. This would take 3.488(75) =261.6 seconds or 4.54 minutes.

This is a good example of a direct relationship and an example of a physical law. Nature does not boil water at different speeds depending on mood, attitudes, or apprehension. Heat and water follow the same rules all the time; however, many of the systems in our social environments do not always follow such well-defined and predictable rules. In our physical environment, we are able to use mathematics to determine exactly what will happen, but in our social environment, we can only use mathematics to model what can or should happen. For example, in economic systems, there is a strong relationship between supply and demand. However, because the system involves human attitudes, choice, and freewill, it is impossible to make the same kinds of accurate predictions about how an increase in price will impact demand, as can be made when determining how long it will take water to boil.

Our social systems are impacted by the human characteristics that influence how we interact with our environment. For this reason, our social systems do not have such hard and fast laws as heating water. Human systems are less predictable. When we boil a liter of water on our stove, it will always take the same amount of time. Human beings do not do anything the same way every time. This does not mean that we cannot predict and manage human systems; it simply means that we must expect and live with a certain amount of error. We can predict exactly how long it will take water to boil, but we cannot accurately predict what the cost of gas will be next week.

Our inability to make accurate prediction in human systems may be a result of our inability to accurately measure all of the cause and effect relationships that exist in a system, or it may be that the cause and effect relationships are totally unpredictable. If the change in component B is always different for a given change in component A, then we may not be able to make accurate predictions at all. Still there is enough predictability in human systems that mathematics is a powerful tool that can be used to measure, predict, and manage change. However, using mathematics to manage your social environment requires that you move from a qualitative perspective to a quantitative perspective by collecting and analyzing data to understand cause and effect relationships in a system of change.

A Quantitative Approach to Dieting and Nutrition

At some point in time, we all think about our weight and our diet. We may notice clothes fitting a little tighter or just an extra pound or two or three. We try to eat less and exercise more to lose weight, but how much more and how much less? We do not usually make accurate measurements of how much more we exercise or how much less we eat. The only measurement we take is our weight. When we weigh ourselves, we are either encouraged or discouraged by the one measurement. Without knowing how much more we exercised, if any, or how much less we have eaten, if any, we have only one way to assess our efforts and results, the scales.

Qualitative, guess and check, approaches to dieting support a billion-dollar industry of diet aids and weight loss programs. There are lots of exercise programs that promise that if you follow their exercise routine you will lose weight. However, many of the exercise programs do not highlight that your success with their program will require dietary changes. That is left for the fine print after you have purchased the program. Other programs emphasize eating a healthy low-calorie diet and downplay the exercise portion of their program. The designers of these programs know that weight loss involves a combination of calories eaten, calories burned during exercise, basic metabolic rates, nutrition, and a whole host of human characteristics that are perhaps most important of all. They have taken a quantitative approach to designing their diet program.

You should take a quantitative approach to understanding their programs. A quantitative approach to dieting places weight loss in the context of a system of nutrition, calories, exercise, and metabolic rates. In this system, you can understand exactly what it takes to lose (or gain) weight. Once you understand this system, then you are in a position manage your weight. Once you understand the system, you can understand how the weight loss systems you see on TV work and which system, if any, will be best for your goals.

A Weight Management System

The calorie is the basic unit of measurement used in discussion of diet, nutrition, and weight loss. When we say that food contains 100 calories, we are say that the body will convert that food into that amount of energy. This is like saying that 1 gallon of gas is equivalent to 25 miles. Technically, one calorie is the amount of energy needed to raise 1 gram of water 1 degree. So if you remember our discussion about boiling water, one calorie is equal to 4.184 joules. In the USA, we actually misuse the term *calorie*. In Europe and other countries that use the metric system, food labels show kilocalories, but in the United States, we use just the word *calorie* to mean the same thing as kilocalories. So on US food labels, one US calorie is equivalent to 1 kilocalorie in other countries.

Your body requires a certain number of calories to maintain body weight. These calories are used to create the energy needed for your heart, lungs, liver, and kidneys to function. This is your basic metabolism rate (BMR). There are several formulas for computing BMR that use more or fewer variables. A rough estimate is that for women this number is about 11 times the body weight, and for men it is about 12 times your body weight. The rest of the calories that you burn in a day are used in physical activity and depends on how physically active you are.

To lose weight, you must burn more calories that you take in. This is a rule of nature. Every weight lose program and or dietary aid is based on getting you to burn more calories than you eat. About 4000 calories equals a pound. Suppose you go to the gym every other day. If, on a gym day, you burn 1000 calories more than you take in, you will lose ¼ of a pound. It will take you four days to lose one pound. If on the next day (the no gym day) you eat the exact same amount, then you will regain the ¼ pound (since you did not burn the extra 1000 calories in the gym), so your net two-day weight loss will be zero. To lose weight by going to the gym every other day, you must eat no more than you BMR on the days that you don't go to the gym. If you eat your BMR on the days that you do not go to the gym, it will take about a week to lose one pound.

How many calories do you eat per day? How many calories do you burn per day? What is your BMR? These are the numbers and measurements that you must control if you want to lose weight. These are the cause and effect relationships that you must understand if you are to be able to set realistic weight loss goals. If you do not know these numbers and or understand the cause and effect relationships that govern weigh loss and gain, then you cannot control them. This leaves you at the mercy of the weight loss industry that makes billions off of your ignorance. In fact, if you knew how many calories your were going to burn in a given day and you knew how many calories you were going to eat, then in order to gain weight you would have to make a conscious decision every day to eat more than you will burn. Thus, perhaps the best weight loss program is just to know and understand the numbers.

The major components in your weight loss system are BMR, calories eaten, and calories burned. We should also add nutritional requirements to the system. You can get 3000 calories by eating a variety of foods that will contain different amounts of minerals and vitamins that the body needs for good health. Some foods are more easily converted to energy, and some provide a quick burst of energy while others provide a long slow supply of energy. Some exercises burn more calories than others. Some exercises build muscle, which may actually lead to weight gain. Thus, we can add food types, nutrition, and exercise types to your system.

Technology can make it easier to measure the amount of calories that you eat each day and the calories that you burn. There are many programs that will compute the number of calories in the foods that you eat. The number of calories per serving is also listed on the package and is available in some restaurants. Smart cell phones can make it easier to count calories. We almost always have the cell phone with us, and there are programs that run on the phones that will track calorie data. Some of these applications will also compute the number of calories that you have burned in a day. Some companies are manufacturing wearable devices that automatically measure and record the number of calories you burn each day.

A quantitative approach to weight control puts you in control of your weight. It does not make it easier to lose weight; it simply allows you to see the real-time effect of your decision to have dessert or not to have dessert. It also allows you to decide to eat only half of the baked potato to save room for the dessert. It gives new meaning to the question—did you save room for dessert? If you know how many calories you have eaten and how many you have burned, then you know whether you have room for dessert, as opposed to judging if you stomach has reached its stretching limits.

Should food be sold by the calorie? Imagine a restaurant where you can specify not only what you want to eat but also how many calories you want. The restaurant would then adjust the size (and price) of your portions based on your calorie level and whether you have chosen to have dessert.

When we add factors like nutritional values, glycemic index, carbohydrates, fats, cholesterol, and the all-important feeling of being full, managing your diet can become more complex. However, shouldn't you know what you are eating and drinking? As you increase the number of relevant variables that you measure in your system, you gain greater control over the system. Taking a quantitative perspective toward your diet is a step toward engaging in a quantitative discourse about diet. It moves you from the grand promises of weight loss to a reasoned and rational process of understanding that allows you to set and manage reasonable goals.

Why should you make the move to a quantitative discussion of your diet? There are health reasons and financial reasons. Weight loss and dieting is billion-dollar industry that produces lots more promises than results. This means that millions of people are purchasing weight loss (weight gain) products and programs that are not getting the results that they desire. Taking a quantitative approach to your diet can significantly reduce the likelihood of being taken advantage of by promises of fast easy weight loss. The other reason is, of course, to improve you health. What two better reasons are there than money and health? I am sure that there is even a connection between a good and healthy diet and peace of mind.

The amount of mathematics that you need or use will depend on the questions that you ask and your desire to know. You may want to know you average daily caloric intake and the average number of calories that you burn each day. The difference between these numbers will give you your average weight loss or gain per day. These numbers will tell you how long it will take to reach your weight loss goals. You may want to measure vitamin and nutritional values in foods as compared to number of calories. You may want to compare the cost of healthy eating to other diets and to minimize your food cost while meeting all of your dietary needs. You may want to maintain charts that record changes in calories eaten and calories burned over months and/or years so that you can identify trends and patterns in your weight loss or gain based on changes in your diet or daily routine. As always, the amount of mathematics that you need will depend on what you want to understand. However, to get started you must take a quantitative perspective and enter into the quantitative discussion about weight control. You do not have to do the mathematics, but it is your money, health, and peace of mind that are at stake.

Taxes: A Complex System

Taxes provide another example of a system that we often discuss qualitatively. We all pay taxes. We often complain about the amount of taxes we pay. We agree or disagree with how our taxes are being spent. Sales taxes, state taxes, federal taxes, property taxes, gas taxes, hotel taxes, liquor taxes—where does all of the money go? How much money is collected in taxes? Are we being overtaxed?

Politicians tell us that they won't raise taxes and then they do. Are our schools getting better? Are our cities cleaner, safer, better environments to live in? We often ask some of the right questions, but we satisfy ourselves with simple qualitative answers like worst, better, too much, or not enough. We need to have a more quantitative discussion about it.

Most people will agree that some government is necessary and so taxes are necessary. When it comes to government, we often have disagreements over what is essential and necessary versus what is expendable and unnecessary. These issues are relevant to how much we pay in taxes because this is what we are buying with our tax dollars. However, we could pay more taxes and get less government, or we could pay less tax and get more government. Before we can determine if we are getting a bargain on government, we need to know how much we pay in taxes.

It is easy to manipulate qualitative discussions with propaganda and scare tactics that are not based in quantitative analysis. If you want to get to the truth, you will have to join a quantitative discussion about taxes. A quantitative discussion about taxes is not based on what is right and wrong. It is based on understanding the numbers. Once you understand the numbers, you will be in a better position to judge right, wrong, necessary, and unnecessary.

I am not offering answers or great insights here, and I am not presenting an argument for one position or another. I am inviting you to participate in a quantitative discussion about your well-being. You should not agree or disagree with what I say; you should question what I say and then go out and find your own answers. When politicians express a need to raise taxes or when they promise a tax cut, we should ask for more information to assess the validity of the politicians' recommendations. Consider the following comments by Johnson, Oliff, and Williams in an article published by the Center on Budget and Policy Priorities.

> The national recession is producing both declines in state and local revenues and increased need for public programs as residents lose jobs, income, and health insurance. In the 2009 and 2010 fiscal years, the imbalance between available revenues and what was needed for services opened up budget gaps in most states. In addition, states have now addressed significant budget shortfalls in enacting their 2011 budgets and even more budget gaps are projected for fiscal year 2012. Since the start of the recession, states have closed over $425 billion in budget shortfalls. Sizable budget gaps are likely to continue for the next several years.
>
> Virtually all states are required to balance their operating budgets each year or biennium. Unlike the federal government, states cannot maintain services during an economic downturn by running a deficit. States had record reserves heading

> into this recession, but those have mostly been drawn down. Since federal economic assistance is slated to expire well before state budgets have recovered, states must address remaining shortfalls with a combination of spending cuts and/or tax increases.

In times like these (above), it seems even more necessary to have quantitative discussions about budgets. If we want things to work, it is our responsibility to hold politicians, government officials, and business leaders responsible. You must understand the system. Understanding how much we pay in taxes is just a place to begin to understand one of many systems that shape and define our social environment.

How Much Do We Pay in Taxes?

How much do we pay in taxes? How much tax is paid by various groups in society? How much tax do corporations pay? How much total tax revenue is collected? To have a quantitative discussion about taxes and to know if we are paying too much in taxes, we must first understand exactly how much we pay in taxes. To provide a measure of how much we pay in taxes, our quantitative discussion begins with a comparison of taxes to the gross domestic product (GDP).

If we know that in 2007 we paid a total of $2.568 trillion dollars in taxes, what do we know? This is certainly a lot of money, but one number by itself does not really tell us much. We need to know how much money we made before we can place any real meaning to this number. For example, suppose that Jim makes $50,000 dollars and pays $6,000 dollars in income taxes and Lisa makes $75,000 and also pays $6,000 in income taxes. While they both paid $6,000, Jim paid 12% of his income in taxes while Lisa paid 8% in taxes. Based on these numbers, we might argue that the situation is not fair for Jim, but we don't really know if Jim or Lisa paid too much in taxes.

The GDP is a measure of the nation's income. It is a measure of how productive the country has been. Comparing how much we pay in taxes as a country to the GDP provides an objective measure of how big $2.568 trillion really is. This comparison gives us an unbiased answer to the question how much do we pay in taxes. In 2007, the GDP was $14.077 trillion dollars. Thus in 2007 we paid 18.24% of GDP in taxes. So 18.24% of our productivity went toward providing funds to run the government and government programs. Is 18.24% of GDP too much? Have we always paid this much, or are we paying more now than in the past?

Historical Comparisons

Historical comparisons of tax revenue to GDP from previous years will tell us if the amount of taxes we are paying has remained relatively stable or if there have been significant changes.

Table 3

RECEIPTS AS PERCENTAGES OF GDP: 1930–2015										
Year	**%**		**Year**	**%**		**Year**	**%**		**Year**	**%**
1930	4.2		1951	16.1		1972	17.6		1992	17.5
1931	3.7		1952	19.0		1973	17.6		1993	17.5
1932	2.8		1953	18.7		1974	18.3		1994	18.0
1933	3.5		1954	18.5		1975	17.9		1995	18.4
1934	4.8		1955	16.5		1976	17.1		1996	18.8
1935	5.2		1956	17.5		TQ	17.7		1997	19.2
1936	5.0		1957	17.7		1977	18.0		1998	19.9
1937	6.1		1958	17.3		1978	18.0		1999	19.8
1938	7.6		1959	16.2		1979	18.5		2000	20.6
1939	7.1		1960	17.8		1980	19.0		2001	19.5
1940	6.8		1961	17.8		1981	19.6		2002	17.6
1941	7.6		1962	17.6		1982	19.2		2003	16.2
1942	10.1		1963	17.8		1983	17.5		2004	16.1
1943	13.3		1964	17.6		1984	17.3		2005	17.3
1944	20.9		1965	17.0		1985	17.7		2006	18.2
1945	20.4		1966	17.3		1986	17.5		2007	18.5
1946	17.7		1967	18.4		1987	18.4		2008	17.5
1947	16.5		1968	17.6		1988	18.2		2009	14.8
1948	16.2		1969	19.7		1989	18.4		2010 estimate	14.8
1949	14.5		1970	19.0		1990	18.0		2011 estimate	16.8
1950	14.4		1971	17.3		1991	17.8		2012 estimate	18.1

The amount of taxes as a percent of GDP was increasing by about 3.8% per year until 1942. Then, in 1942 the Revenue Act was passed. The Revenue Act increased the number of people paying taxes from around 5% to about 75% and introduced payroll tax withholding. Individual tax rates also rose during this time. The amount of taxes paid as a percent of GDP rose from 6.8% in 1940 to 20.4% by the end of the war. After the war, the percentage fell slightly to 14.4% before rising again to 19% in 1952 during the Korean War. The pre-WWII percentage of 7% and 6% were never to be seen again. Between the Korean War and the Vietnam War, the amount of taxes as a percentage of GDP hovered around 17.5% before reaching a high of 19.7% in 1969 during the Vietnam War. We have maintained a wartime percentage every sense, so that when the war in Afghanistan and Iraq began, there was only a slight rise in taxes as a percentage of GDP.

The amount of taxes as a percentage of GDP has gone from 4.2% in 1930 to a high of 20.9% in 1944 and to an average of 17.3% over last ten years. War certainly played a major role in tax increases, but war is not the only factor. Also why have we never been able to reverse the effect of the war tax? Was 4.2% enough? Was 20.9% too much? The world has certainly changed a lot since 1930, so we should probably not expect to operate a government on 4.2% of GDP. But what changed? What do we need to measure to understand the increase in tax as a percent of GDP?

We could measure changes in the number of government employees or changes in the number of government agencies. We could measure changes in social programs like Medicare. If we are going to measure social programs, we should also measure the growth in the number of millionaires that have benefited from being able to do business in the stable system provided by our government. Well, we used to count millionaires; now we might have to count billionaires. We could measure government spending on research and changes in military spending. If we are going to understand why we pay 18.24% of GDP on taxes instead of the 1930 level of 4.2%, we should probably measure all of these things and more. Then we could have a quantitative discussion about how much we pay in taxes that could determine if we pay too much.

International Comparisons

In order to gain a bit more perspective on our tax bill, we can compare the ratio of tax revenues and GDP in the United States to the same ratio in other countries. This will provide a little more perspective on the size of our tax burden.

Table 4

	1975	1985	1990	1995	2000	2004	2005	2006	2007 Provisional
Canada	32	32.5	35.9	35.6	35.6	33.7	33.4	33.3	33.3
Mexico		17	17.3	16.7	18.5	19	19.9	20.6	20.5
United States	25.6	25.6	27.3	27.9	29.9	26.1	27.3	28	28.3
Australia	25.8	28.3	28.5	28.8	31.1	31.1	30.8	30.6	n.a
Japan	20.9	27.4	29.1	26.8	27	26.3	27.4	27.9	n.a
Korea	15.1	16.4	18.9	19.4	23.6	24.6	25.5	26.8	28.7
New Zealand	28.5	31.1	37.4	36.6	33.6	35.3	37.5	36.7	36
Austria	36.7	40.9	39.6	41.2	42.6	42.8	42.1	41.7	41.9
Belgium	39.5	44.4	42	43.6	44.9	44.8	44.8	44.5	44.4
Czech Republic				37.5	35.3	37.8	37.5	36.9	36.4
Denmark 1	38.4	46.1	46.5	48.8	49.4	49	50.7	49.1	48.9
Finland	36.5	39.7	43.5	45.7	47.2	43.4	43.9	43.5	43
France 1	35.4	42.8	42	42.9	44.4	43.5	43.9	44.2	43.6
Germany 2	34.3	36.1	34.8	37.2	37.2	34.8	34.8	35.6	36.2
Greece	19.4	25.5	26.2	28.9	34.1	31.2	31.3	31.3	n.a.
Hungary				41.3	38	37.6	37.2	37.1	39.3
Iceland	30	28.2	30.9	31.2	37.2	38	40.7	41.5	41.4

Ireland	28.7	34.6	33.1	32.5	31.7	30.1	30.6	31.9	32.2
Italy	25.4	33.6	37.8	40.1	42.3	41	40.9	42.1	43.3
Luxembourg	32.8	39.5	35.7	37.1	39.1	37.3	37.8	35.9	36.9
Netherlands4	40.7	42.4	42.9	41.5	39.7	37.3	38.8	39.3	38
Norway	39.2	42.6	41	40.9	42.6	43.3	43.5	43.9	43.4
Poland				36.2	31.6	32.3	32.9	33.5	n.a
Portugal	19.7	25.2	27.7	31.7	34.1	33.9	34.7	35.7	36.6
Slovak Republic					33.8	31.8	31.8	29.8	29.8
Spain 1	18.4	27.6	32.5	32.1	34.2	34.6	35.8	36.6	37.2
Sweden	41.2	47.3	52.2	47.5	51.8	48.7	49.5	49.1	48.2
Switzerland	23.9	25.5	25.8	27.7	30	28.8	29.2	29.6	29.7
Turkey	11.9	11.5	14.9	16.8	24.2	24.1	24.3	24.5	23.7
United Kingdom	35.2	37.6	36.1	34.5	37.1	35.4	36.3	37.1	36.6
Unweighted average:									
OECD Total	29.4	32.7	33.8	34.8	36.1	35.2	35.8	35.9	n.a
OECD America	28.8	25	26.8	26.7	28	26.2	26.9	27.3	27.4
OECD Pacific	22.6	25.8	28.5	27.9	28.8	29.3	30.3	30.5	n.a
OECD Europe	30.9	35.3	36.1	37.1	38.4	37.5	38	38	n.a
EU 19	32.2	37.6	38.2	38.9	39.4	38.3	38.7	38.7	n.a
EU 15	32.2	37.6	38.2	39	40.6	39.2	39.7	39.8	n.a

In 2006 taxes as a percent of GDP in the USA was less than every country except Japan, Korea, Mexico, and Turkey. Are general conditions are better in the United States than in these other countries? This data would suggest that we do not pay too much in taxes. Of course, this international comparison is just the tip of the iceberg. There are many more comparisons that have to be made to understand the differences in national tax rates. Does a dollar buy more or less government than a euro, peso, or a yen? We could also make international comparisons of health care spending, military spending, spending on infrastructure, government payroll, and more. This data provides one more way to begin to contextualize the development of a system in which to understand how much we pay in taxes.

Tax Groups

Once we have a better sense of how much we are paying in taxes, we might decide to look at who is paying the most taxes. If taxes have to go up or go down, why does the percentage that is paid by one group or the other need to change? Should everyone share equally in any changes in demand for taxes? Does everyone profit equally from the benefits provided by a tax increase or decrease?

First we can consider what percentage of tax is paid by individuals, by corporations, from excise taxes, and from other sources. Table 5 shows the amount of tax paid in total by individual taxpayers, by corporations, through excise taxes, and by other sources. Data are listed from 1934 to estimates for 2015. This data gives us an opportunity to see how the distribution of the tax burden has changed over time. The graph below represents the percent of total government tax revenues that came from each of the categories. In the graph, we can see that initially excise and other taxes accounted for more than half (72.5%) of government receipts. By 1945, corporate and individual taxes accounted for 76.1% of government receipts. However, starting in 1946, corporate taxes began to account for less and less of the total government receipts while the percent contributed by individual taxpayers has remained between 40% and 50% of total government receipts. Corporate taxes now account for less that 15% of government revenue.

Individual taxpayers have been asked shoulder more and more of the tax burden while corporations have been free to collect record profits. Note also that a large portion of the excise tax is paid by the individual taxpayers. This situation provides an interesting context in which to consider the rise in the number of corporate millionaires. Of course, if corporations are asked to pay more in taxes, they could eventually pass their cost on to the consumer, especially if the consumers are not paying appropriate attention to the numbers.

Figure 4

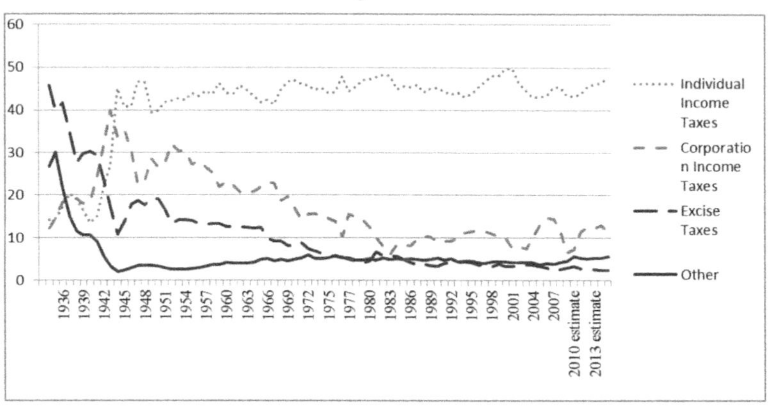

Individual Tax Brackets

How is the tax burden distributed among individual taxpayers? We have a progressive tax system. This means that the more money you make the more taxes you pay. The intent is to shift more of the tax burden onto those who can most afford to pay. The number of tax brackets changes from time to time. In 2010, there were six tax brackets. There were also four filing statuses: married filing joint, married filing single, single, and head of household. Your filing status together with your income help to determine your tax bracket. If you are married filing joint returns or the head of household status with taxable income of $83,000 you would be in the 25% tax bracket. If you are single or married filing separate returns with $83,000 in taxable income, you would be in 28% tax bracket. Apparently single people and those with a second family income can afford to pay more taxes.

In 2010, the lowest individual tax rate was 10% and the highest rate was 35%. Between 1944 and 1963, the highest tax bracket was more than 90%. This is just unreal. That means that if you had a taxable income of $1,000,000, the government collected $900,000 in taxes. Between 1964 and 1980, the highest tax bracket was 70%. This is still unbelievably high. Assuming that people actually paid this much in taxes, this is clearly too much. Are you kidding me? No wonder people were looking for tax shelters and other ways to hide money.

The government provides lots of loopholes, tax shelters, and tax incentives to help hide money. Marriage, children, home ownership, and 401ks are a few incentives available for individual taxpayers. There are also incentives for banks, businesses, and corporations that encourage companies to invest in jobs, particular kinds of research, and activities that create jobs. The list of incentives and loopholes are too large to list.

Over the last 20 years, the highest tax bracket has been on average about 35%; however, everyone is looking for ways to reduce their tax liability by taking advantage of government incentives. Those who make more money are able to take advantage of more incentives and shelters and are also able to hire accountants to hide more of their money thus lowering their tax liability. All of this makes it difficult to untangle who is paying what in taxes.

Summary

There are a number of issues that have to be considered if we are going to have a quantitative discussion about taxes. Taxes as a percent of GDP, international comparisons, tax rates on individuals, corporate tax rates, and excise taxes. The system that we must construct to answer the question—do we pay too much in taxes?—is large and complex. The answers lie in a system of measurements and comparisons.

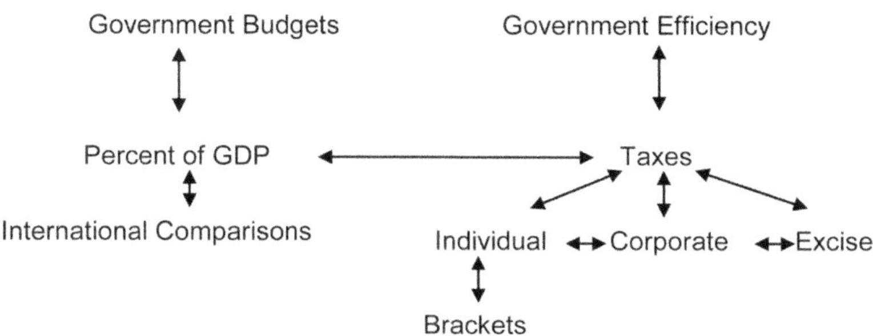

We have not discussed the important topic of government efficiency. Inefficient governments do not necessarily result in higher taxes, and efficient governments do not necessarily lead to lower taxes. However, we certain prefer to buy efficient government. According to the Office of Management and Government, the Obama administration seeks to improve government efficiency by implementing the following three management strategies.

1. **Use Performance Information to Lead, Learn, and Improve Outcomes**. Agency leaders set a few high-priority goals and use constructive data-based reviews to keep their organizations on track to deliver on these objectives.
2. **Communicate Performance Coherently and Concisely for Better Results and Transparency.** The Federal Government will candidly communicate to the public the priorities, problems, and progress of Government programs, explaining the reasons behind past trends, the impact of past actions, and future plans. In addition, agencies will strengthen their capacity to learn from experience and experiment
3. **Strengthen Problem-Solving Networks.** The Federal Government will tap into and encourage practitioner communities, inside and outside Government, to work together to improve outcomes and performance management practices.

These strategies represent an invitation to the American people to pay attention to and to participate in the quantitative assessment of government programs.

> The Government Performance and Results Act of 1993 (GPRA) and the Performance Assessment Rating Tool (PART) reviews increased the production of measurements in many agencies, resulting in the availability of better measures than

> previously existed; however, these initial successes have not lead to increased use. With a few exceptions, Congress does not use the performance goals and measures agencies produce to conduct oversight, agencies do not use them to evaluate effectiveness or drive improvements, and they have not provided meaningful information for the public.

Measurements are being produced and not used. The measurements are there for you. The administration has taken steps to encourage senior officials to make better use of assessment data. You can help with encouragement by joining the quantitative discussion and by expecting to be given access to the reports. You can hold government agencies accountable by joining the quantitative discussion.

> Going forward, agencies will take greater ownership in communicating performance plans and results to key audiences to inform their decisions.

You need to become a part of the key audience. Good data is power. Good data holds the power to confirm success and expose failure. Data is good if it measures the right things and if it can be interpreted within the context of other data. As we have seen, a single measure of taxes collected is data, but not good data. We need more data in order to interpret this number. Thus good data provides enough information to be interpreted.

Understanding the System

The government provides services that make it possible to operate a business, work, and live. The government provides physical infrastructure (roads and bridges), social infrastructure (schools, healthcare), and protection (FDA, FCC, police, military, SEC, etc). These things are necessary.

There are three ways we could be paying too much in taxes. First, we could be paying for services that we are not receiving. Second, we could be paying for services that we don't need. Third, we could be placing too large a burden on the tax base (hitting you too hard in the checkbook), which could lead to problems in the economy.

To prevent the first case, we have to pay more attention to how money is spent. We need to know what we are supposed to be receiving. We need to know the services that are to be provided and the budget, and we need to track government expenditures and assess the level of service. The government has various offices that conduct these analyses. The budget is available. The service goals of departments and agencies are available. We can track government expenditures, and we can track goal completion rates.

The second issue is a bit harder to get at. We can certainly compare the cost of a program to the benefits derived from the program. However, it is sometimes difficult to measure and determine benefits. If a program saves one life, then do we need the program and at what cost? The quantitative analysis will provide an answer, but it may be difficult to accept. It is hard to place a value on human life.

In both cases, it is easier to assess program effectiveness and the need for a program when the resulting benefits can be easily quantified. The third case is just such a situation. The third case helps to balance the first two by placing limits on what we can do. If there was an endless money supply, then we would not need to debate the usefulness of a program, and efficiency would not be a major concern; however, there is a limit to the amount of money that can be collected in taxes without disrupting the very environment that the government is supposed to provide. These and other quantitative discussions have already been started. It is time for your voice to be heard in the discussion.

Table 5

Fiscal year	Total	Individual income taxes	Corporate income taxes[1]	Excise taxes[2]	Other[3]
In million dollars					
1934	2,955	420	364	1,354	788
1935	3,609	527	529	1,439	1,084
1936	3,923	674	719	1,631	847
1937	5,387	1,092	1,038	1,876	801
1938	6,751	1,286	1,287	1,863	773
1939	6,295	1,029	1,127	1,871	675
1940	6,548	892	1,197	1,977	698
1941	8,712	1,314	2,124	2,552	781
1942	14,634	3,263	4,719	3,399	801
1943	24,001	6,505	9,557	4,096	800
1944	43,747	19,705	14,838	4,759	972
1945	45,159	18,372	15,988	6,265	1,083

1946	39,296	16,098	11,883	6,998	1,202
1947	38,514	17,935	8,615	7,211	1,331
1948	41,560	19,315	9,678	7,356	1,461
1949	39,415	15,552	11,192	7,502	1,388
1950	39,443	15,755	10,449	7,550	1,351
1951	51,616	21,616	14,101	8,648	1,578
1952	66,167	27,934	21,226	8,852	1,710
1953	69,608	29,816	21,238	9,877	1,857
1954	69,701	29,542	21,101	9,945	1,905
1955	65,451	28,747	17,861	9,131	1,850
1956	74,587	32,188	20,880	9,929	2,270
1957	79,990	35,620	21,167	10,534	2,672
1958	79,636	34,724	20,074	10,638	2,961
1959	79,249	36,719	17,309	10,578	2,921
1960	92,492	40,715	21,494	11,676	3,923
1961	94,388	41,338	20,954	11,860	3,796
1962	99,676	45,571	20,523	12,534	4,001
1963	106,560	47,588	21,579	13,194	4,395

1964	112,613	48,697	23,493	13,731	4,731
1965	116,817	48,792	25,461	14,570	5,753
1966	130,835	55,446	30,073	13,062	6,708
1967	148,822	61,526	33,971	13,719	6,987
1968	152,973	68,726	28,665	14,079	7,580
1969	186,882	87,249	36,678	15,222	8,718
1970	192,807	90,412	32,829	15,705	9,499
1971	187,139	86,230	26,785	16,614	10,185
1972	207,309	94,737	32,166	15,477	12,355
1973	230,799	103,246	36,153	16,260	12,026
1974	263,224	118,952	38,620	16,844	13,737
1975	279,090	122,386	40,621	16,551	14,998
1976	298,060	131,603	41,409	16,963	17,317
TQ	81,232	38,801	8,460	4,473	4,279
1977	355,559	157,626	54,892	17,548	19,008
1978	399,561	180,988	59,952	18,376	19,278
1979	463,302	217,841	65,677	18,745	22,101
1980	517,112	244,069	64,600	24,329	26,311

1981	599,272	285,917	61,137	40,839	28,659
1982	617,766	297,744	49,207	36,311	33,006
1983	600,562	288,938	37,022	35,300	30,309
1984	666,438	298,415	56,893	37,361	34,392
1985	734,037	334,531	61,331	35,992	37,020
1986	769,155	348,959	63,143	32,919	40,233
1987	854,288	392,557	83,926	32,457	42,029
1988	909,238	401,181	94,508	35,227	43,987
1989	991,105	445,690	103,291	34,386	48,321
1990	1,031,972	466,884	93,507	35,345	56,188
1991	1,054,996	467,827	98,086	42,402	50,665
1992	1,091,223	475,964	100,270	45,569	55,731
1993	1,154,341	509,680	117,520	48,057	50,783
1994	1,258,579	543,055	140,385	55,225	58,440
1995	1,351,801	590,244	157,004	57,484	62,596
1996	1,453,055	656,417	171,824	54,014	61,386
1997	1,579,240	737,466	182,293	56,924	63,186
1998	1,721,733	828,586	188,677	57,673	74,966

1999	1,827,459	879,480	184,680	70,414	81,052
2000	2,025,198	1,004,462	207,289	68,865	91,730
2001	1,991,142	994,339	151,075	66,232	85,529
2002	1,853,149	858,345	148,044	66,989	79,011
2003	1,782,321	793,699	131,778	67,524	76,342
2004	1,880,126	808,959	189,371	69,855	78,534
2005	2,153,625	927,222	278,282	73,094	80,902
2006	2,406,876	1,043,908	353,915	73,961	97,271
2007	2,568,001	1,163,472	370,243	65,069	99,610
2008	2,523,999	1,145,747	304,346	67,334	106,417
2009	2,104,995	915,308	138,229	62,483	98,058
2010 estimate	2,165,119	935,771	156,741	73,204	123,647
2011 estimate	2,567,181	1,121,296	296,902	74,288	139,579
2012 estimate	2,926,400	1,326,045	366,361	81,085	148,002
2013 estimate	3,188,115	1,468,410	393,474	84,994	171,001
2014 estimate	3,455,451	1,603,861	444,805	86,503	188,077
2015 estimate	3,633,679	1,733,476	411,055	87,829	206,681

Chapter Five

Systems Analysis

If mathematics is truly all around us, then why do we make up word problems? Is it because we can't find a connection to the students' lives? Perhaps we are not the ones that should be making the connection.

So far we have relied heavily on the use percent and percent change in our efforts to understand change. You can develop a good deal of understanding with these simple tools. Perhaps, more importantly you can use percent and percent change to identify questions and issues about change that need to be investigated. Still, you may be wondering where the probability, statistics, algebra, and the calculus are and why you would need to know more than percent and percent change. The answer to this question should be getting clearer to you. We have not needed statistics, probability, algebra, and calculus because we have not yet asked any questions that require us to use these concepts.

Remember this book is not about teaching you mathematics, but about getting you to ask the kinds of questions that require mathematics. The goal is to get you to ask the kinds of questions that you need in order to understand the systems that make up your social environment. If you do not develop this understanding, then you will not be able to manage your environment. You will not be able to be a responsible participant in quantitative discussions that will determine your future.

Statistics, probability, algebra, and calculus are used to perform higher-level analysis of change in systems. For very well-behaved systems, we have algebra, analysis, and calculus, and for less well-behaved systems, we may rely more heavily on statistics and probability. More advanced levels of mathematics are for developing more advanced levels of system understanding. The initial understandings that you begin with (and that we have developed in the book so far) about the system are simple understandings of quantitative change, based on and represented by percent and percent change. As you understand the systems, you also understand percent and percent change. To understand one is to understand the other. Asking more detailed questions about health care, investing, diet, or taxes will lead to both the need to use more mathematics and to an improved ability to understand the mathematics. The questions will help you understand mathematics because the mathematics will be a representation of the system changes that you are trying to understand. The system and the mathematics become one in the same.

The understanding process begins with a review of the data that we have collected. We look for patterns in the changes and for relationships between the changes in two or more variables. In the series of measurements of national health care expenditures, we can see that the expenditures are increasing every year. We are able to measure the size of the change using percent change, but percent and percent change alone can only tell us so much. Percent change tells us how big the changes are relative to and initial value. Another characteristic of change that is important is the speed of the change. The data in table 6 or in any table do not "move." To see change, we need to represent the movement of the data from one value to the next by graphing the data. We want to know how the percent change is changing.

Table 6

National health expenditure
In millions

Year		Year		Year	
1960	27,534	1976	152,478	1992	849,039
1961	29,370	1977	172,826	1993	912,485
1962	32,053	1978	194,126	1994	962,061
1963	34,910	1979	219,940	1995	1,016,271
1964	38,694	1980	253,373	1996	1,068,526
1965	42,173	1981	293,592	1997	1,124,915
1966	46,430	1982	330,743	1998	1,190,059
1967	52,062	1983	364,676	1999	1,265,158
1968	59,012	1984	401,599	2000	1,353,187
1969	66,396	1985	439,284	2001	1,469,359
1970	74,894	1986	471,265	2002	1,602,284
1971	83,265	1987	512,973	2003	1,734,932
1972	92,974	1988	574,043	2004	1,854,840
1973	103,034	1989	638,794	2005	1,980,603
1974	116,809	1990	714,127	2006	2,112,668
1975	133,124	1991	781,608	2007	2,241,208

In figure 4 we see two similar graphs of the data from table 4.

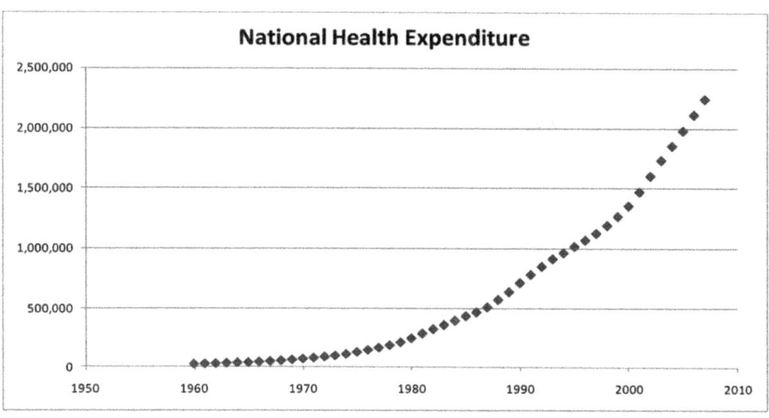

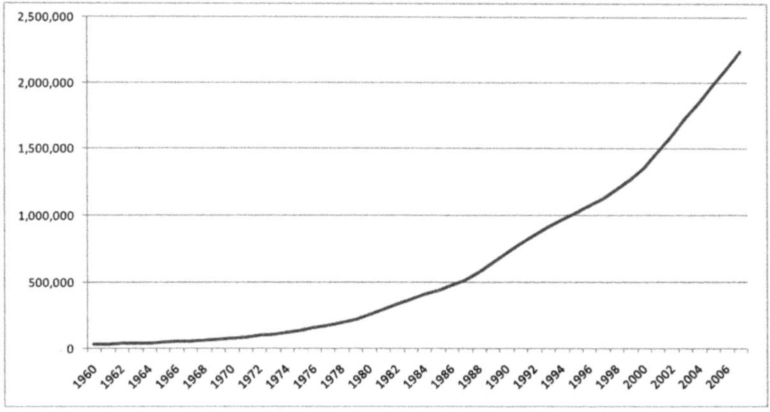

Figure 4

In the graph on the left, we see each individual measurement is represented by a dot. In the graph on the right, the same data are represented as a continuous line. The graphs help us to understand a bit more about how the data is changing. We already knew that spending was increasing. In the graphs, this is represented by the upward turn as we move from left to right. However, in the graphs, we can see that not only are the values increasing, but we can also see that the change is speeding up. The changing in speed is represented by the increased steepness of the curve. Every bend in the curve represents a change in the speed at which health care expenditures are increasing. But what causes the turn or bend in the graph?

This change in speed is the central problem in health care. Now, we should expect the cost of health care to rise just as the price of everything else rises; however, the cost of health care is rising faster than the cost of everything else. This is a central quantitative issue that needs to be discussed with respect to health care reform.

How fast are costs changing? How fast should they be changing? When we computed percent change in the previous chapters, we were able to answer the question how big is the change. For example, we can see that a 6% change is not as big as a 14% change. Now we want to understand how fast health care is changing. To do this, we need to compute the change in health care spending over a specific amount of time. From the table 6, we can see that in 1962 health care expenditures were $32, 053 and in 1965 spending was $42,173. We can use these numbers to compute the average rate, or speed, of change over the period 1962 to 1965. If you subtract the amount spent in 1962 from the amount spent in 1965, and then divide that number by the number of years from 1962 to

1965 (i.e., 1965–1962) the final answer will be the average rate of growth in spending from 1962 to 1965.

$$\frac{1965 spending - 1962 spending}{196501962} = \frac{42,173 - 32,053}{1965 - 1962} = \frac{15020}{3} = \frac{5006.6}{1}$$

This says that spending was increasing at a rate of $5 billion a year over the three-year span from 1962 to 1965. If spending had continued to increase at this rate, then the graphs in figure 4 would be straight lines. However, as we will see by computing the rate of change over different time periods, health care spending did not continue to grow at this rate.

Let's compare the growth rate in 1965 to the rate of change in other time periods. Again, using the data in table 6, the average rate of change from 1967 to 1970 is given by the equation,

$$\frac{74,894 - 52,062}{1970 - 1967} = \frac{22832.}{3} = \frac{7610.6}{1}.$$

Here you can see that the rate of change increased to more than $7.5 billion dollars a year. So healthcare spending was growing 2.5 billion dollars a year faster from 1967 to 1970 than it did from 1962 to 1965. Why did the growth speed up? I don't know the answer to that question. Do you? Let's keep going. The average rate of change from 1977 to 1980 was

$$\frac{253{,}373-172{,}826}{1980-1977}=\frac{80{,}547}{3}=\frac{26849}{1}$$

Wow! Here again you can see that the rate of change increased. Health care spending was growing almost 20 billion dollars a year faster in 1980 than it was in 1970. You should stop and think about this for a minute. But wait—it gets worst. Let's jump ahead a bit and see how things looked in 2000. The average rate of change from 1997 to 2000 was

$$\frac{1{,}353{,}187-1{,}124{,}958}{2000-1997}=\frac{228{,}272}{3}=\frac{76{,}090}{1}$$

Holy cow! Health care was growing 50 billion dollars a year faster in 2000 than it was in 1980. You should really stop and think about this. In 1970, health care was growing at a rate of 7.6 billion dollars a year. By 2000 it was growing ten times faster, at 76 billion dollars a year. Holy $#+!@%! Why on earth is health care spending growing that fast? I believe that this is a direct result of the fact that far too many people are not engaged in the kinds of quantitative discussions that would have led to this question being asked back in 1980. Just imagine how much health care will cost by 2010.

I selected four different periods of time at random to measure the rate of change in health care spending. What you see is that the rate of change has increased from $5 billion dollars per year in 1965 to $76 billion dollars per year in 2000. It is this increase that causes the line in figure 4 to curve. The line curves up significantly because the rate of change has increased significantly. This is the difference between driving the family minivan and driving a Corvette ZR1. In 1965, the rate of change in health care spending was like driving a minivan. By the year 2000, it was as if we had traded in the minivan for a much faster ride, with fewer seats.

A rate of change is more than just a difference in growth from one year to the next or over several years. In our system, we use the rate of change to determine how fast things are changing and where they will be in the future. We are able to ask the question, what would the health care spending level be if spending continued to grow at the rate of growth experienced in 2000?

If the rate of change did not change after 2000, then we would expect that in 2003 we would be spending $1,581,457 in health care expenditures, but we actually spent $1,734,932. The reason that we underestimated the amount spent in 2003 is because the change is not constant. The change is speeding up. There are two questions. First, why is health care spending increasing, and second, why is the increase in health care spending increasing so much faster year after year? Each of our calculations above gives us the speed of change over a specific period of time. We can see that the speed of the change is getting larger, that is, that the change is accelerating.

In chapter 2, we compared heath care spending to population growth to see if population growth could help to explain the rise in healthcare spending. We determined that that the percent change in population growth was not a good predictor of the change in health care spending. Figure 5 is a graph of population data shown in table 7.

Table 7 US population in millions 1960 to 2007

Year		Year		Year	
1960	186	1976	222	1992	260
1961	189	1977	224	1993	263
1962	192	1978	226	1994	266
1963	195	1979	228	1995	269
1964	197	1980	230	1996	271
1965	200	1981	233	1997	274
1966	202	1982	235	1998	277
1967	204	1983	237	1999	280
1968	206	1984	239	2000	283
1969	208	1985	242	2001	285
1970	210	1986	244	2002	288
1971	213	1987	246	2003	291
1972	215	1988	248	2004	294
1973	217	1989	251	2005	296
1974	218	1990	254	2006	299
1975	220	1991	257	2007	302

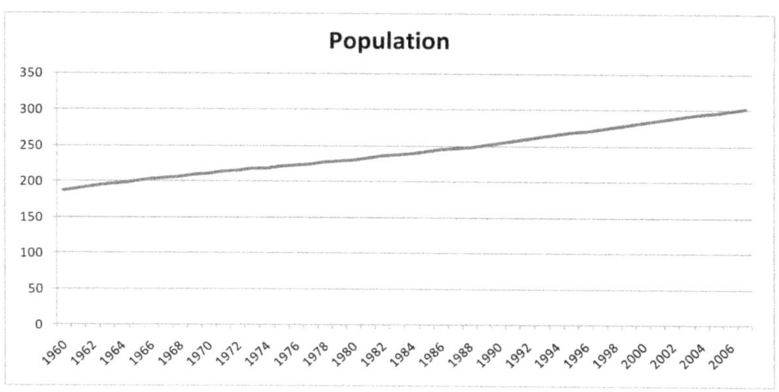

Figure 5 US population in millions

What we see in the data in table 7 is that the population grows by roughly the same amount each year. That is why the graph of the data in figure 5 is almost a straight line. The graph in figure 4 curves up because the rate of change is increasing. Let us compute the average change in population for a few time periods to see what we can see.

$$\frac{200-192}{1965-1962}=\frac{8}{3}=\frac{2.66}{1}, \quad \frac{210-204}{1970-1967}=\frac{6}{3}=\frac{2}{1},$$

$$\frac{230-224}{1980-1977}=\frac{6}{3}=\frac{2}{1}, \quad \frac{283-274}{2000-1997}=\frac{9}{3}=\frac{3}{1}$$

In 1965, the population was growing at a rate of 2.66 million people per year. In 1970 and in 1980, the rate had decreased to 2 million per year until the year 2000 when the rate increased to 3 million people per year. We can see that the rate of change in the population has remained relatively constant at between two and three million people per year. There has not been the kind of acceleration that we see in the graph of the health care data.

The average rate of change that we have computed in the last two examples is the slope of the line or the curve. Just like the slope of a hill tells us how fast the hill rises or falls, so to the slope of a line tells us how fast the data is changing. The slope provides a quantitative way to discuss the speed at which the data is changing.

Every change in the slope is a change in the speed. All change is caused by something. Thus we can use a change in the slope to identify interesting points in time to investigate in order to develop greater understandings of the cause and effect relationships in our systems. For example, we can see that in the health care data there is a significant change in speed in 1970 and throughout the seventies. We might want to look to see if there is anything that happened in that time frame that might help to explain the rise in spending. There were changes in Medicare around that time.

Note that in both examples the graphs show growth; however the population grows like a slow steady drip while the health care spending grows like a leak in a dam that gets bigger and bigger with each passing moment.

Remember that variables may have a direct relationship, or they may have an inverse relationship. In either case, we would expect that two variables that have some kind of relationship between them to show similar changes in their graphs. When we compare the two graphs, we see that there is nothing in the population data that seems to explain the changes in health care spending. We might expect to see a significant change in population around 1970; however, we do not. This would suggest that population growth is not the major cause of the rise in health care spending.

We might look at other system variables that have similar change patterns that might help to explain the rise in healthcare spending. When we compare spending in various areas of health care, we see that they are not all increasing in the same way. There are two areas that are increasing much faster than the others. Hospital care costs and physician and clinical services costs are rising much faster than other health care areas. The cost of home health care, dental services, and other professional services rise at a rate that is more consistent with population growth and inflation. Thus we need to consider factors that are causing hospital care and physician and clinical service to rise so quickly.

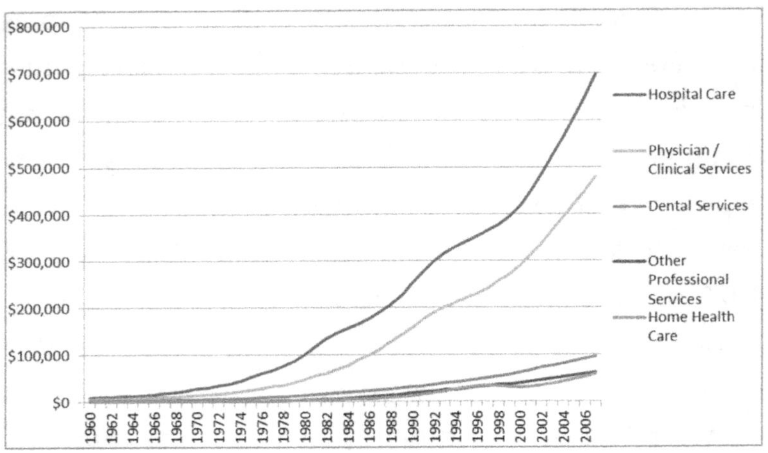

Figure 6

Here we can see an important point about mathematics. Mathematics can point to questions, but it does not provide explanations. Mathematics provides the quantitative foundation for making comparisons and decisions; however, it is up to the individual to interpret the mathematical results within a system of understanding. If the system of understanding is faulty, then so too may be the application of the mathematical results. We need a much better understanding of hospitals and physician services to be able to understand what might be driving up the cost in these areas.

Linear Versus Nonlinear Change

The difference in the graphs of the population data versus the health care spending data are helpful in understanding two important mathematical concepts. The change in population data is almost linear while the change in the health care spending data is clearly nonlinear. Linear and nonlinear change are key concepts that are necessary for the development of quantitative understanding. These concepts are central to the question, how is the change changing. These concepts are important in your efforts to understand changes in your social environment.

Linear and nonlinear are important ways to describe change. It is necessary to understand these two concepts in terms of change in system variables and not to think of them as abstract mathematical concepts. Linear and nonlinear are quantitative descriptions of how things change that allow us to replace qualitative terms like fast and slow with more precise measurements of rates of change.

The change in population is roughly the same in any given year. However, the amount of change in health care spending changes every year. One year it changes by 5 billion, and in a different year it changes by 21 billion. How fast the health care spending is changing depends on when you ask the question. Since population grows at the same rate each year, to compute the population at any point in time you can start by selecting 1960 as the starting year and then you simply multiply 2.5 million times the number of years since 1960 and add the result to the starting value in 1960. For example, in 1960 the population of the United States was reported as 186 million people. To approximate the population in 1970, you multiply 2.5 times 10 (the number of years since 1960) and then add the result to the initial population from 1960.

$$\begin{aligned} &1970 \text{ population} &&= 2.5 \text{ times } 10 + 1960 \text{ population} \\ \text{Or } &1970 \text{ population} &&= 2.5(10) + 186 \\ & &&= 25 + 186 \\ & &&= 211 \text{ million people} \end{aligned}$$

In table 7, we can see that the population was reported as 210 million people. Our equation overestimated the actual number because the population does not really change at 2.5 million people every year. In figure 7, we can see a comparison of the graph of our estimated population using our equation to the graph of the actual data. We see that, in fact, our formula will overestimate some values and underestimate others; however, it gives a good approximation. The data is not exactly on the line because the change is not exactly the same each year. Thus every change in the speed of population growth (faster or slower) moves us off of our line (above or below) that represents a constant change of 2.5 million people per year. Constant change is like driving in your car with the cruise control set at 50 mph. Constant change is predicable. If you drive 50 mph, then you predict where you will be at each point in your trip. However, if your speed changes up and down (and so is not constant), it is more difficult to predict where you will be in 2 hours. Most of our daily estimates of change involve an assumption of constant or some average rate of change. We are familiar and comfortable with constant change, and we have incorporated it into our qualitative discussions.

In order to engage in quantitative discussions and to understand the quantitative changes in your social environment, you will need measure and quantify change that is constant and change that is not constant. Constant change and nonconstant change are referred to as linear and nonlinear respectively. Constant change is called linear change because the graph of constant change looks like a straight line. Nonconstant change is called nonlinear change because the graph of the change does not look like a straight line.

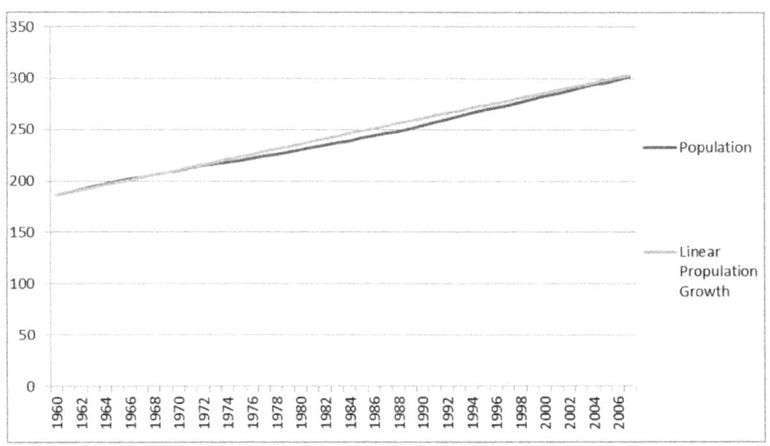

Figure 7

Generating Equations

Having an equation to approximate the data can simplify many aspects of the effort to understand change. Previously, we computed an estimate of the 1970 population using the formula

1970 population = 2.5 times 10 + 1960 population.

Since the change (2.5 per year) is the same every year, this same equation can be used to estimate the population for any year after 1960. Let's rewrite the equation using variables to make in shorter and more general.

1970 population = 2.5 times 10 + 1960 population

$p = 2.5$(years since 1960) $+ 186$ (To compute population for any year after 1960)

$p = 2.5(x) + 186$ (It is easier to write x than to write years since 1960)

This last equation can be used to compute an estimate for any year after 1960, and as we saw from our graph in figure 7, the estimate is relatively accurate. Thus, the equation can be used in place of the data in many situations, and if you want to know what the population will be in 2020, we can estimate using the equation.

$p = 2.5(60) + 186$ or $p = 150 + 186$ so $p = 336$

So based on our equation, we estimate that the population will be 336 million in the year 2020. This only works because the average change from year to year is the same, 2.5 million people per year.

It would be helpful if we had a similar equation for predicting health care cost. We can compute the average change in health care spending, but since the average change depends on the year, you cannot use the average change in 1970 to predict the spending in 1990. To determine how fast the health care data is growing, we must employ more advanced mathematical thinking and tools. However, the idea is fairly straightforward.

You start by generalizing the equation for the population data. We had that $p = 2.5(x) + 186$. This equation is for a constant change of 2.5 per year and for a starting value of 186. We could generate a similar equation for any situation where the change was constant and based on some starting point. Thus the equation $y = mx + b$, where m is the constant change and b is the starting value, is a general equation for computing linear data. Remember that linear is a kind of change. We know that the change in health care spending is changing. Now let's assume that the way that it is changing is linear.

Now an equation like $y = mx + b$ has a graph that is a straight line just like the population data. A straight line to represent the change in health care spending would look like y = average change in health care spending (# of years) + starting value.

A straight line would not be a good way to approximate health care spending (see figure 6). The reason is that the rate of change or **m** in a straight line does not change, but when **m** is the average change in health care spending, **m** does change. If we start with an equation for a line as a bad approximation for the way that healthcare data is changing and then we assume that **m** (the change) is changing linearly, then we get the following equation.

y = change in health care spending (x) + starting value

becomes

y = (nx + b)(x) + starting value

Do not try to understand this last equation in terms of the mathematics that it represents. You must understand it in terms of what it says about the change in health care spending. It is not the mathematics that is getting more complicated; it is the questions that we are asking about health care spending. Your drive to understand the last equation most likely depends more on your desire to understand health care spending that on some need to understand mathematics.

Remember our question. Why is the change in health care spending rising so fast? Well, the last equation says that the change in the change in health care spending is linear. Now, since I know a bit of mathematics, I know that the change in the change in healthcare spending is not really linear, but for now we will use the assumption to make a point.

We can use our knowledge of mathematics to help us determine an equation that can generate the health care data. This is where general knowledge of mathematics separated from how you use it is important. If we multiply $y=(nx+c)x+b$, we get $y = nx^2+cx+b$, which is a quadratic equation. If we know what the graph of a quadratic equation looks like and how we can manipulate the graph, then we can determine a quadratic equation that may more closely approximate the graph of the actual health care spending data. This is the mathematics that you learn in algebra.

The point is that if you want to analyze changes in the speed of health care spending, then you are beginning to ask the kinds of questions that will require more advanced probability, statistics, algebra, and calculus. You may not know this mathematics right now, so this discussion may be more or less clear to you. However, if you understand the question that we are trying answer about health care spending (why is the change in health care spending changing so fast), then you are closer to understanding the mathematics than you think. If you understand constant speed and the constant slope and you understand that a change in speed means a change in slope, then it is a short step to see that we are just looking for an equation to represent how the speed is changing.

If you are getting my point here, then you are not trying to learn the mathematics. Instead you are trying to determine how fast health care spending is rising at any point in time. If you want to know and understand health care, then you are probably still reading with interest. However, if this has become a purely mathematical exercise for you, then you may have missed the question and the point of the discussion.

I won't pursue the discussion of finding an equation to represent the change in speed in health care spending any further for now. I will point out that it is at this point that having general knowledge of mathematics becomes important. At this point in the attempt to understand change in the health care spending, it would be helpful to be able to apply some generalized understandings of change. You could develop this generalized knowledge of change over time by your frequent engagement in quantitative discussions. However, while you do need to engage in lots of quantitative discussions, developing the kinds of generalizations and insights that are needed would take a lot of time. At this point, the power of mathematics to provide generalized models and representations of quantitative change becomes important. Instead of having to develop these generalizations on our own, we can apply the mathematics concepts that are generalizations of scientists' and researchers' efforts to understand change. These are the concepts that you study in the mathematics classroom, and this is why you have to know them. Or the fact that you do not want to understand health care spending (or anything else) is why you do not need to know mathematics. It's all about trying to understand change.

We will end this discussion for now with a graph in figure 8 of the change in the speed of growth in health care spending. This is a graph of the change in the change in health care spending. Notice that it is not linear. The change in the change does not change the same way all the time. What we can see is that there are periods of slow growth and periods where the growth was decelerating. Note that this does not mean that spending was going down—it just means that it was not speeding up as fast. There are also periods of significant acceleration. In order to understand health care spending, you must understand what causes these changes in the speed of spending growth. One consequence of more people asking these kinds of questions might be that more of the answers would be provided. Do you need to understand health care spending?

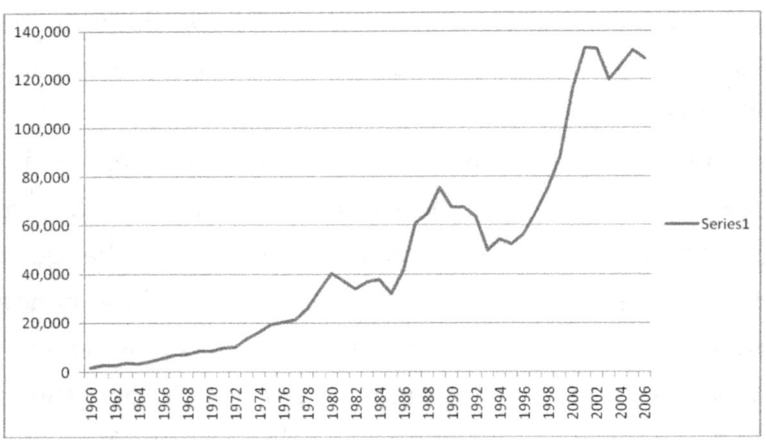

Figure 8

We can see that graphing the data provides a visual representation of the change in the data that helps us to analyze changes. Knowing how fast or how slow things change is the key to being able to predict change. When we can determine how fast something is changing and how the speed is changing, then we are able to make predictions about where it will be and about how fast it will be changing in the future. When we can predict these things, then we can develop equations that allow us to determine future values.

Our predictions are based on the assumption that whatever is causing the thing to change will continue in the same way. That is why we compare changes in different variables in the system so that we can identify things that seem to cause our variable of interest to change. We compared changes in population growth and changes in health care spending; however, there did not seem to be much of a relationship between the two. Although they both increase in value, there is no change in the population data that seems to correspond to the significant increases in the change in speed at which health care spending is growing.

Mathematical equations are the tools that we use to record and represent the cause and effect relationships that exist between variables in a system. Mathematical concepts and their properties are like the laws of change. They are very much like the laws of physics. The laws of physics come from our attempts to understand our physical environment. The laws of mathematics come from our much-related efforts to understand change in our physical and social environments.

Tax Data Analysis

At this point, you should be beginning to understand that the essential question here is not whether you need to know mathematics but whether you need to understand health care spending. I believe that there is only one correct answer for a responsible American citizen. But if health care spending does not catch your interest, then perhaps you would be interested in understanding the environment, or your weight and health, or the economy. There are a number of key issues that you need to understand to manage and control your own future and to be an informed, participating citizen who helps to manage and control the future of our society.

Let's look at one more example that we have already discussed. We can apply graphing and the ideas of linear and nonlinear change to the questions we raised about taxes. Why should you or shouldn't you expect a tax cut? Whether or not we need a tax break depends on the roles that taxes play in our economic system. First, taxes keep the economic system working. Taxes are used to create and enforce laws that make "fair" business dealing possible. They provide infrastructure needed to move goods, people, and information that are essential for the operation of businesses. So no reasonable person would suggest that taxes are not necessary; however, a reasonable person may question the efficiency of government spending and may questions whether the distribution of the tax burden is spread in a way that optimizes cost and benefits for all citizens.

The question of government efficiency is difficult to measure. The White House and state governments provide goals and objectives for various government departments and offices that can be used to measure efficiency. This makes it possible to determine if goals and objectives are being met, but reaching a goal and competing an objective does not ensure efficiency. One may even question whether the goals and objectives that have been set are the most efficient goals and objectives. Local and national governments have access to and use a great deal of data when identifying goals and objectives; however, just how this data is used is not always clearly communicated. Government decisions are not always justified by quantitative arguments. They are often justified by qualitative arguments. This is in part because politicians and government officials are not held accountable by you to provide the quantitative data, the systems, or the data analyses that they used to set goals and objectives. This is the quantitative discussion that we must all join with our elected government officials. This is not intended to suggest that all decisions should be based on quantitative analysis alone; however, it is probably the case that no decisions should be made without considering the data.

Any further analysis or discussion of government efficiency would take us into the realm of political opinion, an area that I have tried to avoid. Also there is insufficient data easily available to me at this time to make starting and/or demonstrating such an analysis possible. However, the most important issue has been accomplished, I hope. That is, I hope that you are beginning to see the need to address the quantitative issues that are at the core of many of your social systems. I hope that you are beginning to understand the need to engage in quantitative discussions about the education system, the health care system, the communication system, the insurance system, the banking system, and many other systems that directly impact your life.

The second quantitative issue related to taxes concerns the way that taxes are distributed. This discussion will start with a consideration of the information presented in figure 9.

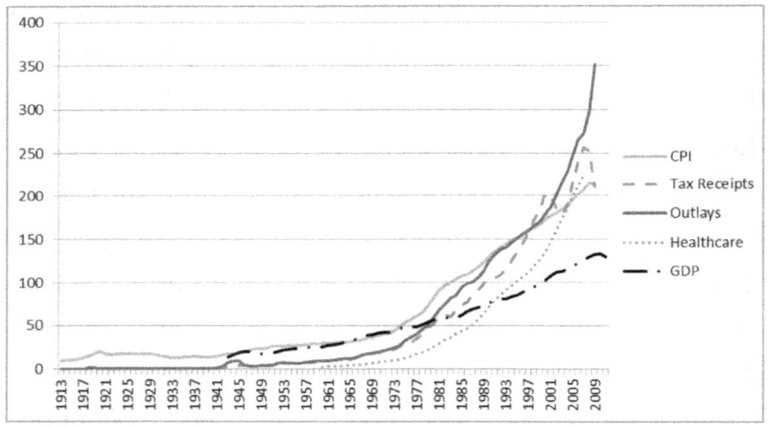

Figure 9

Figure 9 shows the growth in the consumer price index (CPI), tax receipts, government spending (outlays), health care, and the gross domestic product (GDP). Note that the CPI is measured in hundreds while other variables are measured in millions, billions, or trillions. Thus all of these variables have been scaled so that they can be displayed on the same graph. In figure 9, we can see that at one point government spending and tax revenues were balanced. As government spending went up, taxes went up.

However, during the 1970s, government spending began to outpace tax revenues. In the 1990s, we see that tax revenues surpassed spending, and then in the 2000s, we see a return to spending exceeding revenues. The graph also shows that health care has risen at a similar rate as has government spending

This is what we can see from a quick glance at the graph. A more detailed analysis of the data would allow us to identify changes in the rate of change in the various variables. This might help to understand the relationship between changes in outlays and the GDP, or allow us to understand how changes in tax revenues are related to the changes in the CPI. There are clear changes in tax revenues that occurred in the 1990s and the 2000s during the Clinton and Bush presidential administrations, respectively. However, only a detailed analysis of how the changes are changing will tell the whole story and lead to a greater understanding of the system.

Our expectations for a tax increase or decrease must be understood in the context of a complex economic system. The graphs provide a way to see some of the key relationships that need to be investigated. However, the role played by your desire to understand this system cannot be overstated.

Data collection and observation are the starting point. In addition to data, there are significant historical events like wars and natural disasters, government policies, and other factors that need to be considered in terms of their impact on the system. (What data and analysis went into the Clinton and Bush decisions?) Mathematics is for measuring, predicting, and managing change. There is perhaps no system that people try to manage and manipulate more than the economic system.

Government rules exist to protect the innocent and uniformed consumer. Let the buyer beware, but the buyer is not aware, so the government steps in to protect you. The system is so complex that it is almost impossible for the buyer to be aware. We do not have the resources or the time to understand safety issues relating to home furnaces, airplane engines, microwave ovens, or cell phones. However, we cannot and should not expect the government to do it all without our participation. The more aware we are, the less government has to do for us and the more we will recognize the need for government efforts to manage our economic system. The more we understand, then the better prepared we are to participate in the management of our social systems.

What is the impact of a tax reduction or an increase in government spending? If consumer spending increases, do prices go up? How do we stop prices from rising? Perhaps if we have well-informed buyers, then prices could be controlled. We could consider the NFL and other sports leagues as an example of how smart buyers control prices. In the NFL, owners, smart and informed as they may be, continue to drive up the price that they pay in salaries to top athletes. By the way this price tag is passed on to you. Where do you think that companies get the three or four million dollars to run a 30-second commercial during the Super Bowl? They pass that price directly to the consumer. Television stations pay enormous amounts of money for the exclusive rights to broadcast sporting events and then pass their cost on to the advertisers, who in turn pass the expense on to you. Broadcasting companies make their money from advertisements and viewer subscriptions, so again it all comes back to you and me because we purchase the game. Making smart buying decisions is more complicated than it might seem.

How does a rise in taxes affect the GNP and the CPI? How does government spending affect jobs and interest rates? You may be waiting for me to do a lot of analysis to answer these questions, but instead I hope that you and others will pick up the charge. My answers would be based on my own assumptions, beliefs, views, and ideas. This book is not about what I think about these issues. This book is about getting you think more and more quantitatively about these issues. If you are left asking questions, then this book may have had the desired impact. If you are asking quantitative questions, then the book is closer to its mark. And if you are measuring change, indentifying cause effect relationships, and investigating the relationships between changes in your system, then the book has hit its mark.

Why Do I Have to Know Mathematics?

Whether it is economics, investing, taxes, diet and nutrition, health care, or the environment, everything changes. You need to be able to measure, predict, and manage change in order to make the best personal and civic decisions. There are millions of questions that you need to be asking. You need to join the quantitative discussion.

Average rainfall

Average temperature

Shopping is more than a great sale

The value of education

Crime rates in your community

International economics

Housing prices

Gas prices/energy cost

Your health

Poverty

Technology

I suppose that I could make this list as long as you want it, but in the end your questions are what are important.

As I have been writing this book, I have been torn between applying lots of mathematics to provide fully developed examples of mathematical analysis and keeping my focus on motivating you to ask questions. For the most part, we have only used percents in our analysis. In the current chapter, we began to use the concepts of linearity, nonlinearity, slopes, and rates of change. These ideas are actually the entry point for the use of more advanced algebra and calculus concepts. Again an effort on my part to demonstrate these concepts would take me to far afield from the major point of the book. The need for this level of mathematics depends on you starting to ask questions that require much less complex mathematics. You have to walk before you run.

When we have enough system knowledge, then we can begin to develop our understanding of how the system is changing. How is the change changing? More advanced levels of mathematics are for developing more advanced understanding of systems. So if you need or want to have more advanced understanding of systems, then you will ask more advanced questions about how things are changing. These questions will lead to both the need to know and the ability to understand mathematics. The questions will help you understand mathematics because it will be a quantitative representation of the system. The understandings that you begin with about the system then help you develop better mathematical representations, all the while helping you to understand the mathematics. Before long not only will you need to know mathematics, but you will find that learning mathematics has taken on a whole new meaning.

Moving Forward

If I have been successful in helping you to understand why you need to know mathematics, then you also know that you need to begin to participate in more quantitative discussions about change in your environment. I have purposefully not used topics like sports, fashion, stamp collecting, or gambling as topics. You may be interested in these and other topics and hobbies; however, these are not the things that you *need* to understand. Instead, I picked topics like health care, taxes, diet, and the economy because these are things that have a direct impact on the quality of your life so you *need* to understand them to be able to manage your life.

My argument is that if you begin to engage in more quantitative discussions about change in your environment, then you will also begin to need, learn, and use more mathematics. You might wonder if you need to know more mathematics before you can begin to engage in more quantitative discussions. This is not a chicken or egg kind of situation. The quest for quantitative understanding definitely came and can come before the development of mathematics. So the answer is that you do not need to know more mathematics to start your quest to understand your environment. In fact, the whole point is that your quest to understand your environment will lead you to learn more mathematics.

In our school systems, we have reversed the situation by attempting to teach mathematics when there is no quest for understanding. The fact that students are not using mathematics to reason about and understand their environment may be a major reason that students struggle in mathematics classrooms. This is the reason that so many students ask the question, "Why do I have to know mathematics?" Students have been telling teachers, schools, and curriculum developers what the real problem is in the mathematics classroom since the first student asked this question. So why don't students use mathematics to reason about their environment?

If you have or if you know young children, then you know that they ask lots of questions. Children are curious about their environment. They want to know where wind comes from and why the sky is blue. They want to know how doorknobs work and where babies come from. Children ask lots of questions. This seems like the perfect setup for teaching mathematics. However, there are at least two things that go wrong with this initial setup.

First, when little Johnny and Mary get to the first grade and meet their teacher, they immediately start to ask questions. Why does the bell ring? Why can't I fly like a bird? Where does wind come from? The typical teacher response to all of these questions is to tell the students that they must sit quietly and raise their hands when they want to ask a question. So Mary runs to here seat and raises her hand, and when called on she asks, "Where do stars go during the day?" Her teacher's reply is usually something along the lines of "You have to stay focused on what we are talking about in class?" Being young, impressionable, and eager to learn, Mary and Johnny do exactly as the teacher asks.

By the fourth grade, most students have learned that when you go to school you do not get to ask questions about what you are interested in learning about. You have to be interested in what the school wants to teach you. Since children are still young and impressionable at this age, they have become good at raising their hands only to ask questions about what the teacher is teaching. Then they enter puberty. Their views of the world and of themselves begin to change, and they have a million new questions. However, they already know that they cannot ask these questions in school. They begin to question the reason for and the relevance of everything. They are no longer so young and impressionable, and they begin to place greater value on their questions than on what the teacher is talking about. The result is that they stop asking questions altogether.

By the time the students enter the ninth grade, they can see the light at the end of the tunnel. Their primary goal at this point is to become seniors and to get out of school. They are experts at listening to teachers while thinking about the things that really interest them. At this point, they have only one question: "Is it going to be on the test?" At this point, the once young curious students that asked a million questions and had to be strapped to their chairs have learned to sit down, be quite, and wait for someone to tell them what they need to know.

Thus, the first thing that goes wrong in our initial setup is that schools teach students to stop asking questions. By the time the students begin to study advanced topics in mathematics, they have also learned to stop asking questions. A few students may have been able to deal with the internal conflicts over the relevance of schooling and may have been able to retain some glimmer of curiosity in the classroom. Others simply buy into the threats of joblessness or the inability to get into the college of their choice. But for far too many students, by the ninth grade the excitement and drive to ask questions and learn has been suppressed for too long and so deep that they may never get it back.

The second thing that goes wrong with the initial setup for teaching and learning mathematics is directly related to the way that we deal with students' questions in schools. Because we stop students from asking their own questions in school, we also miss the opportunity to teach them how to structure their questions and how to answer their own questions. We miss the opportunity to help students structure their qualitative discussions and help them begin the development of quantitative discussions.

The result is that when we try to teach students the most powerful resource we have for thinking and reasoning about change, they are not engaged in or interested in discussing change.

Imagine a school system where children were encouraged to ask questions. Suppose that we nurtured and developed the curiosity that children bring to school by encouraging them to collect data, develop systems, and analyze the quantitative nature of the changes that occur in their systems. Envision a school system that actively engaged students in quantitative discussions about change. Remember that everything changes and ultimately every job has something to do with change. Thus a school we would be preparing students for active productive lives.

What would all of those quantitative discussions mean for teaching and learning mathematics? Additionally, just imagine the increased level of comprehension and abilities to develop understanding that students would acquire. This new school would be an exciting learning environment driven by student interest. But then how would we teach students what we have decided they need to know? One answer to this question is that if what we are trying to teach students is not helping them to do the things that they want and need to do, then why are we teaching them?

The key is to ensure that students are actively engaged in asking and answering questions about change.

Is there some minimal amount of mathematics that we all need to know in this new school or in life? This is perhaps not the right way to look at things. Is there a minimal list of words that I need to know? I think that in the case of mathematics as in the case of learning words, what I need to know depends on what I am doing and where I am doing it. Also using words leads to the need to learn more words, and so using mathematics will lead to the need to learn more mathematics. Thus our concern should not be about what a minimum content list would look like. Our concern should be focused on the students' ability to use mathematics to develop greater understanding of change.

If we think about teaching mathematics as a means to understand change, then we can rethink and restructure the mathematics curriculum. Do I still need to learn algebra in this new school system? Yes. Because mathematics is the language of change and algebra is to mathematics as grammar is to English. But you need to learn algebra in a different way. You need to know how to use algebra to identify and manipulate relationships in the changes that you measure. If you are always using mathematics to develop understanding, then you will always know why you are learning algebra, because you will be using algebra to measure, predict, and manage change. Let's take a quick look at some of the curriculum in this new school of the study of change.

Arithmetic

First, in order to understand and discuss change, we need to be able to count and to measure. This means that students need to have a solid foundation in the number system, whole numbers, fraction, decimals, and with addition, multiplication, subtraction, and division. This is fairly consistent with traditional learning goals for elementary classrooms. Students still need a strong foundation in the basics. This is like phonics in spelling and reading. Students need to learn the basic tools for discussing change. There are differences between current classroom practice and what I believe has to happen in the classroom for students to learn to embrace mathematics as a valuable resource for thinking. The major difference is that students must spend much more time actually discussing, analyzing, and understanding change in their environment. Students need to count and measure real changes in their physical and social environments as they are introduced to the formal structures, systems, and organizations that surround them. Students must be taught to use mathematics as a tool for reasoning about change in their environment.

In our current system, mathematics has been isolated to a fifty-minute block of time when we teach counting, measuring, adding, and subtracting. We do not then spend the rest of the day showing students what to count measure, add, and subtract. When students learn to read and write in English and language arts classes, they then spend the rest of the day applying these skills in social studies, science, history, and geography classes. They should also apply what they have learned in the mathematics classroom. This may require a different approach to classroom instruction in social studies, science, history, and geography classes.

In the school of change, we correct the situation; however, the correction does not take place in the mathematics classroom as you might expect. Instead the correction takes place in the history, social studies, English, geography, and science classrooms.

We must increase the attention given to quantitative discussions as students learn about the structures, systems, and organization that make up our physical and social environments. Students must learn to use quantitative questions and analysis to help increase their understanding. For example, in the first grade students study topics like *neighborhoods and communities around the world.* Here are four objectives taken from a first grade course in North Carolina. Students will be able to do the following:

1. Describe the roles of individuals in the family
2. Identify various groups to which individuals and families belong
3. Compare and contrast similarities and differences among individuals and families
4. Explore the benefits of diversity in the United States

All of these objects can be accomplish without the use of mathematics by focusing on qualitative discussions of difference. These discussions might focus on physical differences and on qualitative issues like more and less. However, there are also quantitative issues that can be explored to increase students' understanding of how individuals, families, and groups are similar and different.

The goal of such a quantitative discussion would be twofold. First, the goal is to increase the students' understanding of group similarities and differences. Second is to provide an opportunity for students to use what they have learned in the mathematics classroom. This is critical in the early grades as students are beginning to develop their quantitative discourses. Some of the quantitative issues that might be discussed in the neighborhoods unit include the following list:

1. What are the age differences between family members, and are they consistent across groups and generations?
2. What are the weight and height differences among family members and different groups?
3. How many people are there in the United States?
4. How many families of a certain size are in the United States?
5. How many people of various ethnicities, gender, religion, SES, and other categories are there in the United States?
6. How do these numbers changed over time?

You may be wondering how far first graders would get in analyzing the changes in population characteristics. However, in the first grade, perhaps it is enough to get the ball rolling by establishing the importance of considering quantitative questions and having quantitative discussions. Remember, everything changes, so there is always an opportunity to measure, predict, and manage change. Understanding something includes understanding how that something changes. Another way to say this is that there is always an opportunity to engage students in a quantitative discussion about change that will help them to develop greater understanding of whatever the topic may be.

Quantitative discussions and the development of quantitative understanding must be a part of the normal activities in school learning outside the mathematics classroom at all grade levels. Having these quantitative discussions needs to be the norm rather than an occasional addition to a lesson. Students must learn to consider the quantitative aspects of their physical and social environments. It is not enough to teach a child what a percent change is; the child must also learn how, when, and where to use percent change to develop greater understanding of his or her environment. It is also not enough to teach students about static systems, structures, and organization when these systems are dynamic and always changing. Students must learn to use mathematics to help them read and understand change.

We want and expect students to apply what they have learned in the English classroom in all of their other courses. We should also expect students to use what they have learned in the mathematics classroom in all of their courses. This is how students will develop the disposition to engage in quantitative discussions about changes in their physical and social environments. Why don't we engage in these types of discussion on a regular basis? Why aren't there quantitative learning objectives relevant to the understanding of change in all subjects? The development of a quantitative discourse must be a major objective and outcome for elementary grade levels.

Algebra

Perhaps arithmetic is to mathematics what phonics is to reading, and algebra is to mathematics what grammar is to English. Both arithmetic and algebra are essential to the development of a proper quantitative discourse. If students learn to use arithmetic as they engage in quantitative discussions in elementary school, then they will be in a much better position to learn and understand the role that algebra plays in increasing their ability to understand change. They will not need to ask why mathematics is important. They will already know.

Arithmetic is used for measuring absolute change and relative change. Algebra is where we begin to learn how to recognize and characterize patterns of change. As we have seen in other chapters, it is not enough to understand how big a changes is, we also need to understand how and why things change. Understanding how things change means that we must analyze change as a part of a system of measurements and changes. Students study systems in history, social studies, geography, and natural sciences. This means that history, social studies, geography, and natural science classrooms must engage students in quantitative discussions about how systems change.

The foundation for these discussions should have been laid in elementary school. This does not mean that students have to be taught or learn mathematics in the social studies or geography classroom. It means that students should be taught and must learn to develop quantitative understandings in social studies and geography. In junior high school and high school, these courses lay the foundation for how students will approach issues like taxes and health care spending later in their adult lives. If they are not given an opportunity to develop an appreciation for quantitative understanding in junior high and high school, then they probably will not apply quantitative reasoning to these issues in their adult lives.

Also if students are not engaged in quantitative discussions early in history, social studies, biology, and other courses, they will not be ready to apply algebra to the study of how things change. They will not begin to develop the systems needed to contextualize change. But a major reason for learning algebra is so that they can develop mathematical representations of the changes that take place in systems. Thus if students are not engaged in quantitative discussion about systems, then they will begin to ask why they have to learn mathematics. If they do not engage in quantitative discussions, then the one reasonable answer to their question is so that they can get into college, where apparently mathematics will be important.

If on the other hand students are engaged in quantitative discussions, then as they begin to expand their questions and understanding of how things change, they will develop a need for a basic symbol system and language for representing and discussing change. As students continue to expand their understanding of the quantitative nature of change, they will be able to see and understand the use of mathematical models to represent change in systems. They will understand and improve their ability to reproduce patterns of change in their physical and social environments as mathematical equations. Students will develop an understanding of how to interpret the properties and results of mathematical models and computations in the context of the physical and social realities that they represent. They will learn to use mathematics to develop understanding and communicate their understanding to others.

Do I need to be able to factor a polynomial? Do I need to know what a factor is? Do I need to know what at polynomial is? The answer is yes, but not in the way typically defined in mathematics classrooms and texts. You will need to know what a factor is in a system. For example, when studying the economy, we can consider the inflation factor. What does it mean to factor inflation out of or into a quantitative discussion about taxes or health care spending? The mathematical representation of the quantitative discussion might contain polynomials and a variable that represents the inflation factor. Factoring inflation out of the discussion is the same as factoring it out of the equation. So the answer is yes we need to understand polynomials, factors, and factoring. In fact, we need much more than just the simple definitions and procedures that we currently learn in the mathematics classroom. But we do not need more mathematics instruction; what we need is to engage in more quantitative discussions.

Mathematics is for understanding change. If we are not using mathematics in the history, social studies, geography, and science classroom, then the conclusion would be that we are not attempting to develop quantitative understandings of change in these classrooms. What would replace the attempt to develop quantitative understanding of change might be, "Sit down, be quiet, and I will tell you what you need to know."

If we do not challenge students to engineer knowledge from data and facts, then we must provide them with stored facts, names, dates, definitions, etc. We are not training them to be knowledge engineers. However, to be successful in the new global economy we all need to be knowledge engineers. We all need to be able to analyze the data.

In the middle grades, students' understandings of systems, structures, and organization is continued from elementary grades. Students are expected to develop more sophisticated understandings of larger systems. Family structure is replaced by social structures. Students are ready for more advanced quantitative discussions. There are many more opportunities to engage in quantitative discussions, and there is a greater need for these discussions in order to build greater understanding.

At this level, quantitative discussions may need to spill a bit more into the mathematics classroom. In the mathematics classroom, we should focus more on knowing what a factor is in the real world and not just the abstract mathematical definition. Having both understandings is far superior to having just one or the other. It is important to understand that I am not suggesting that any topics be replaced or eliminated in the mathematics classroom or in other classrooms. Instead I am recommending a different way of knowing. I am suggesting that by emphasizing quantitative understanding of systems and system changes, we will both increase students' understanding of content in history, social studies, geography, and science, and we will also greatly improve students' understand and ability to use mathematics. I do not think that anything must be removed from the curriculum, and in many ways, nothing is being added. We are simply taking a different approach. Instead of telling students to sit down and be quiet while the teacher tells them what to know, we are encouraging students to stand up, ask questions, and engage in qualitative and quantitative discussions.

Calculus

If students arrive at the study of calculus with a firm understanding and rich experience using arithmetic and algebra to develop understandings of change in systems, then calculus will make perfect sense. Perhaps nowhere else in the mathematics curriculum is the relationship between mathematics and change so evident. Many calculus textbook authors are quick to point out that calculus is the language and science of change.

As a language, calculus needs to be spoken. It is not clear how much of the science of calculus needs to be learned to use calculus as a resource for understanding. Everyone needs to know what a limit, a derivative, and an integral are in the context of a quantitative discussion about change. It is not, however, the case that everyone needs to know five or six different methods to compute a derivative or integral. This is perhaps where students would take different paths depending on their career choices.

A major issue with respect to what to teach in the mathematics classroom is how to distinguish (or not to distinguish) between students who will eventually learn more advanced mathematics in college and those who will not. I believe that all students will benefit from a greater understanding of the relationship between mathematics and the study of change. Those students who eventually go on to study higher levels of mathematics will be stronger as a result of such understandings. However, here I am most concerned with the teaching and learning of students who typically do not go on to study advanced mathematics in the traditional sense. When these students use mathematics to study and understand change, they will actually be engaging in the same kinds of activities that led to the development of mathematics. They will be using mathematics to learn, not just learning mathematics. In no small sense, they will be mathematicians. As a result, they will be more inclined to study more mathematics.

Probability and Statistics

I have not discussed the role of probability and statistics in the development of quantitative understanding. Probability and statistic do play a major role in quantitative discussions and in developing understandings of system change. We are not always able to determine exact system inputs and outcomes. When a new policy is put in place by the government to address health care spending, or when we start a new diet, it is difficult to say with certainty what the outcomes will be. The concepts and procedures in probability and statistics are used to deal with situations where this kind of uncertainty is present.

Given the uncertain nature of things like human behavior and the economy, it is critical that we incorporate probability and statistics into our quantitative discourse skill set. The inherent uncertainty of our social systems is the reason that statistical analysis is a major component of government and business decision-making processes. You need to understand and participate in this decision process. As with other areas of mathematics, this means that you must begin to ask questions that require probability and statistics. As you begin to understand the questions, then you will also be in a better position to learn and to understand probability and statistics.

Quantitative Discussions in College

As a college mathematics professor, I have sometimes questioned the relevance of mathematics requirements in other majors. Given that students are often told that they need to know mathematics to get into college, I would expect that the need for mathematics in college should be self-evident. I have not done a definitive study on the matter, but based on a review of textbooks in other disciplines, students are frequently not required to use mathematics to do much more that to measure a few variables and perhaps to compute a value or two. Also given the level of mathematics required to satisfy general education requirements and given the way that mathematics is currently taught, it is hard to believe many students are prepared to have any kind of significant quantitative discussions I am advocating that we all should be having.

I am sure that there are counter examples to my claims of low quantitative requirements. However, a primary indicator that my claims hold some validity is that college students frequently do not ask any more questions in their classes than do high school students. Like students in the K–12 system, college students are told what they need to know. If there is a situation when they are required to use mathematics, then they are told which formulas to apply and when to apply them. College students are as likely as anyone else to ask why they have to know mathematics.

Employers often complain that they need graduates who are better prepared to solve problems. Thus, educators design a series of problems for students to solve and use them to teach problem solving. However, far too often students are still told what the problem is and how to solve the problem. Thus the most important step in the problem-solving process is bypassed. The most important step in problem solving and the key to being a better problem solver is asking questions and asking better questions. If students have been taught to stop asking questions as I have suggested, then they will not be good at solving problems. The key to correcting this issue is to teach the students how to ask their own questions. In our current education system, asking one's own questions is a privilege usually reserved for students at the PhD level.

Media and Meaningless Numbers

There are many implications for schools, curriculum materials, and teaching associated with an effort to develop a quantitative discourse. I have tried to touch on some of them above. Ultimately though, the key is still how you interact with the quantitative data, ideas, and issues outside the mathematics classroom. Your primary source of these interactions is through the media.

What do you get when you read the newspaper, watch the evening news, or watch a Sunday morning political talk show? This is where you and I interface with the information generating systems of our adult lives. Do we get detailed quantitative analysis to help support news commentary? During the oil spill crisis in the Gulf of Mexico, it was reported that five thousand barrels of oil were spilling into the gulf each day. What does it mean when you are told that five thousand barrels of oil a day are spilling into the gulf? While I have some sense of the size of a barrel, I do not have the same sense of the size of the Gulf of Mexico. Is five thousand barrels a day like spilling an ink pen into an Olympic sized pool, or is it like a bucket of ink in a pool? I am not trying to minimize the situation, just pointing out that reporting 5000 barrels a day is not providing information.

We are frequently given these kinds of meaningless numbers by our news reporting sources. This is often a clear attempt to strike an emotional response. It is not an attempt to provide useful information.

Media needs to be more accountable for what they report. The only way this will happen is if you and I hold them accountable by participating in the quantitative discussions that are at the core of every issue facing our society today. TV news programs cover everything except the data. Debate shows debate everything except the data. Why is there a program that gives the numbers? The reason is that we do ask for it. Publish the repair records for cars. Give me the Carfax.

Frequently you can find your own information by going straight to the source. If you cannot, then we need to hold the sources accountable for providing the data and information that you need. The information should be in a format that is useful to average citizen. For example, the government provides a great deal of data for public use. Some of this data is presented in easy-to-use spreadsheets, tables, and reports; however, sometimes you need a PhD in information systems and five years of time to understand how to get the data you want.

Conclusion

I hope that you have gotten out of this book what I intended. If you are asking more quantitative questions about your environment, then I have been successful. If you are not asking more questions, then why not?

The fact that mathematics is for understanding change is not a new idea. Many calculus textbooks have noted that calculus is the language of change. Since the purpose of much of high school mathematics is preparation for calculus, the fact that calculus is the language of change might suggest that mathematics instruction should embrace the study of change as a core value.

The problem is not in the mathematics classroom; it is in the history classroom, the biology classroom, the English classroom, and the chemistry classroom. What does it mean if mathematics is not used as a resource for understanding in these classrooms? If mathematics is indeed for measuring predicting and managing change but if it is not used in these other content classrooms for this purpose, then this must mean that students are not measuring, predicting, and managing change in these classrooms. They are not learning how to understand what happened in history from a quantitative perspective. They are not considering the quantitative issues in biology and chemistry. They are not learning how to engineer knowledge.

Why Do I Have to Know Mathematics?

We do not increase our vocabulary for no reason. We increase our vocabulary so that we can communicate better. Learning the word teaches us something about our environment. Learning properties of mathematical concepts should teach us something about how things change in our environment. The usefulness of the word depends on the kinds of conversations that I participate in. The usefulness of the mathematics depends on the kinds of questions that I ask.

I can have an experience that embodies the meaning of a word before I even know the word. Likewise with mathematics, I can engage in an analysis that embodies mathematics concepts before I even know what the concepts are. As I have the experience or conduct the investigation, learning new words or new mathematics concepts is not the goal. Thus, perhaps we learn the most new words and the most mathematics when words and mathematics are not the goal. We learn the most when we are trying to understand and communicate about our understanding.

Instead of learning mathematics and then figuring out how and where to use it, the idea is to figure out what you want to know and understand and then to learn and use the mathematics you need to accomplish your goals. What kind of change would you like to understand? We must be smart consumers of mathematics and statistical analysis. We must expect for the analysis to be produced and reported in a way that we can understand.

So now that you know why you have to know mathematics, the question I have for you is "Do you have to know mathematics?" If you want to understand the quantitative aspects of your social and physical environments, then the answer is yes you need to know mathematics. If the answer is no, you do not need to know mathematics, then you have chosen not to understand your social and physical environment. In the latter case, you have chosen to continue to allow others to manage your life for their best interest.

I hope you need to know mathematics.

Useful Websites

The following is a brief annotated list of websites used in the writing of this book. These websites provide information, reports, and data that will support your efforts to continue the quantitative discussions we have started in the book.

http://www.whitehouse.gov/omb

The Office of Management and Budget

This is a very good place to start your search for information on many important issues. You will find links to the federal budget, analytical documents and historical data. You will also be able to find measurable goals and objectives for government agencies at this site.

https://www.cms.gov/NationalHealthExpendData/

CMS Center for Medicare and Medicaid Services

Historical spending measures annual health spending in the U.S. by type of service delivered (hospital care, physician services, nursing home care, etc.) and source of funding for those services (private health insurance, Medicare, Medicaid, out-of-pocket spending, etc.).

http://www.commerce.gov/

The Commerce Department's mission is to help make American businesses more innovative at home and more competitive abroad. The development of commerce to provide new opportunities was the central goal at the department's beginning in 1903 and it remains a primary obligation today.Comprised of 12 different agencies responsible for everything from weather forecasts to patent protection, the Commerce Department touches the lives of Americans every day.

You will find data on all measures of economic development at this site.

http://stats.bls.gov/

The Bureau of Labor Statistics (BLS) is the principal fact-finding agency for the Federal Government in the broad field of labor economics and statistics. The BLS is an independent national statistical agency that collects, processes, analyzes, and disseminates essential statistical data to the American public, the U.S. Congress, other Federal agencies, State and local governments, business, and labor. The BLS also serves as a statistical resource to the Department of Labor.

BLS data must satisfy a number of criteria, including relevance to current social and economic issues, timeliness in reflecting today's rapidly changing economic conditions, accuracy and consistently high statistical quality, and impartiality in both subject matter and presentation.

http://www.census.gov/
The Census Bureau serves as the leading source of quality data about the nation's people and economy. We honor privacy, protect confidentiality, share our expertise globally, and conduct our work openly. We are guided on this mission by our strong and capable workforce, our readiness to innovate, and our abiding commitment to our customers.

http://www.bea.gov/
The Bureau of Economic Analysis (BEA) promotes a better understanding of the U.S. economy by providing the most timely, relevant, and accurate economic accounts data in an objective and cost-effective manner.

http://www.eda.gov/
This year, the U.S. Economic Development Administration (EDA) marks 45 years of public service, with a mission of leading the federal economic development agenda by promoting competitiveness and preparing American regions for growth and success in the worldwide economy. EDA is an agency within the U.S. Department of Commerce that part-

ners with distressed communities throughout the United States to foster job creation, collaboration and innovation.

http://www.esa.doc.gov/
The Economics and Statistics Administration (ESA) plays three key roles within the Department of Commerce (DOC). ESA provides timely economic analysis, disseminates national economic indicators, and oversees the U.S. Census Bureau (Census) and the Bureau of Economic Analysis (BEA). In this latter role, ESA works closely with the leadership at BEA and Census on high priority management, budget, employment, and risk management issues, integrating the work of these agencies with the priorities and requirements of the Department of Commerce and other government entities.

ESA's expert economists and analysts produce in-depth reports, fact sheets, and briefings on policy issues and current economic events. DOC and White House policymakers rely on these tools, as do American businesses, state and local governments, and news organizations around the world. See our Reports section for a list of recent reports.

http://www.noaa.gov/
NOAA is an agency that enriches life through science. Our reach goes from the surface of the sun to the depths of the

ocean floor as we work to keep citizens informed of the changing environment around them.

From daily weather forecasts, severe storm warnings and climate monitoring to fisheries management, coastal restoration and supporting marine commerce, NOAA's products and services support economic vitality and affect more than one-third of America's gross domestic product. NOAA's dedicated scientists use cutting-edge research and high-tech instrumentation to provide citizens, planners, emergency managers and other decision makers with reliable information they need when they need it.

http://www.cdc.gov/

CDC's Mission is to collaborate to create the expertise, information, and tools that people and communities need to protect their health – through health promotion, prevention of disease, injury and disability, and preparedness for new health threats.

The CDC offers an alphabetical listing of statistics on topics of public health importance, an annual report on trends in health statistics, and much more .

http://www.cdc.gov/nchs/

The National Center for Health Statistics' website, is a rich source of information about America's health. As the

Nation's principal health statistics agency, we compile statistical information to guide actions and policies to improve the health of our people. We are a unique public resource for health information - a critical element of public health and health policy.

www.ingramcontent.com/pod-product-compliance
Lightning Source LLC
Chambersburg PA
CBHW071416170526
45165CB00001B/293

9781456588083